Knowing the Unknown - II
Mysteries of the Universe - Past, Present, and Future

Manohar Lal, Ph.D.

Library of Congress Catalog Number: 201091473
ISBN: 978-0-9826809-1-9

MRLT, LLC, Tulsa
Printed in the United States of America

Table of Contents

ACKNOWLEDGMENTS

Mysteries of the Universe is the second book in the series entitled **Knowing the Unknown**. It addresses the question, *'Where am I?'* It is a question that all of us ask frequently. The first book in this series, *Mysteries of Life*, addresses the question, *'Who am I?'* The third book in the series, **Challenges of Technology**, addresses the question, *'What am I doing?'*

A book of this nature obviously draws upon various sources. Besides some new suggestions, this book is based on my research, papers, articles, and personal discussions with scientists. I want to gratefully acknowledge the help I've received from all sources, which are too numerous to list. In the book, I have listed the names of many distinguished scientists, who have contributed directly or indirectly in understanding the mysteries of science and the Universe. With the popularity of the internet, one can easily find more information about their work by typing a few words into a search engine.

I also want to acknowledge the help of my daughter-in-law, Ritu, who designed the cover page. It would not have been possible to write this book if my friends, my children and grandchildren, and my wife had not inspired me to write it. I dedicate this book to my wife, Rani,who has been my main inspiration in life.

Preface

In this second book in the series on *Knowing the Unknown*, we address the question, *'Where am I?'* We embark on an exciting journey that unfolds a panoramic view of science and the Universe. This book provides a unique perspective on human origins, historical evolution, current frontiers, and the future of science. We will trace back through the history of science to reveal the mysteries of science and the Universe. While searching for the ultimate truth, we will highlight the top ten mysteries of science. This book is meant for readers with some interest in science.

In recent years, several distinguished scientists have written numerous books, expressing their viewpoints. This is, however, not just another book on science. This book integrates the field of science, and brings together the most recent research and observations. It takes an objective look at all the scientific theories. The book raises questions and explores scientific frontiers. We also predict the future of science, and search for the ultimate truth.

This book addresses the questions that most of us have wondered at one time or another. Where did the Universe come from? Where did we come from? What would happen to us and to the Universe? Where do the laws of nature come from? This book should inspire readers and heighten their scientific interest. On reading this book, readers would come to appreciate the elegance of science. It should motivate and encourage young scientists to seek answers to the unanswered questions. This book can also serve as a textbook on the history of science.

An interesting feature of this book is that it explains most of the ideas in simple terms, without invoking complex mathematics. Einstein, the greatest scientist of the twentieth century, once said, "You do not really understand something unless you can explain it to your grandmother." He explained and even conceived his ideas on relativity in terms of simple thought experiments. Of course, mathematics is a powerful tool and an integral part of physics. For mathematically inclined readers, we do give a few equations at the end of the book. These equations give readers a flavor of what scientists have to deal with to carry their ideas to a logical conclusion.

We have organized the subject matter of this book as follows. We start the journey at the point when man first appeared on this planet. At the very outset, we see how science and religion originate. As the journey through science continues, we cross various milestones. Newton explains motion and gravity; Maxwell explains electromagnetism and light waves; Einstein gives the world his theories of relativity, changing the notions of time, space, and gravity. Planck defines the quanta of energy, and Schrödinger introduces quantum mechanics. Then, we visit the entire Universe - its origin, evolution, stars, galaxies, and black holes.

During the journey, we come across the unsolved mysteries of science. We discover several questions that remain unanswered. We discuss these, and prepare a list of the top ten mysteries of science. Towards the end of our journey, we summarize the essence of science and visit its frontiers. We get a glimpse of the future of science, and of some of the exciting discoveries that lie ahead.

As we complete the journey, we search for the ultimate answer. We seek the answer to Einstein's famous question regarding the role of God in creating the Universe. We discuss some fresh ideas on the origin and evolution of the Universe. On the completion of our journey, we would have gained a better understanding and appreciation for the capabilities and limitations of science.

What motivated me to write this book? Mainly, it was due to my personal curiosity to understand the Universe. My unique perspective essentially comes from my diverse educational and professional background. I received an MS and Ph.D. degree in electrical engineering in 1960-63 from the University of Illinois, Champaign-Urbana, Illinois. I taught and conducted research in science and technology for over forty years at several universities and at an Industrial Research Center. This experience provided many opportunities for research in diverse areas of science and technology.

Bon Voyage!

Manohar Lal

Chapter 1
Unfolding the Mysteries of Science

1.1 *Introduction*

Modern civilization owes its existence to science and technology. Both these fields are woven into the fabric of our daily life. As soon as we are born, we start interacting with the Universe through our senses. We feel matter through touch and sense energy through light, heat, and sound. We become aware of space as we move around. We sense time as we sleep at night and wake up in the morning. We observe motion as we chase various objects. A child senses change when we change his diapers. Understanding science is important, since it deals with all these elements: matter, energy, space, time, motion, information, and change. Science also provides rational explanations for most of the phenomena in our Universe.

What is the mission of science? Science attempts to unveil the mysteries of the Universe. It attempts to explain the origin and the existence of the Universe and life. Physical science addresses the question of, 'Where am I?' A scientist, while observing a phenomenon, asks the question, 'Why is it so?' Technology addresses the question, 'How can I control my environment?' It seeks the application of science to make our life more 'comfortable'. A technologist wants to know, 'What can I do?

1.2 *Starting the Journey – An Overview*

Since the dawn of civilization, we have wondered about the origin of the Universe and life. Let us step back in time, and start our journey from the time when man first appeared on this planet. Imagine his emotions and reactions – shock, awe, fear, curiosity, etc. - as he looks around. Obviously, man at this stage is confused, uncomfortable, and afraid of the unknown. His brain, at this stage, cannot find rational explanations for any of the phenomena of nature. He looks at the sky and wonders about the clouds, rain, Moon, Sun, and the other stars. He looks at the surface of the Earth and wonders about the environment – trees, rivers, mountains, and animals. He wonders:

- Where does the rain come from?
- Where did the trees, rivers, mountains come from?
- Where did different animals come from?
- Where did the Moon, sun, and stars come from?
- Where did the Universe come from?
- What is going to happen to the Universe?

- Birth of Science

While some people fantasize about gods, others put their faith in the belief that every phenomenon must have a cause. Science is thus born, and its practitioners are called scientists. They ask why nature behaves the way it does, and why things happen the way they do. Of course, things keep on happening in the Universe whether they understand, or even have the terms to describe, the cause. Scientists continue to march forward and want to know more. They do not want to accept the simplistic explanation that God did everything. They dare not question God because of the fear of retribution, but they do want to know - how did God do it? Unfolding the mysteries of the Universe becomes the mission of science.

Scientists observe various phenomena to find a 'scientific' explanation. They noticed that change is the main characteristic of every phenomenon. It is eternal, perpetual, and immortal. They also observed that nothing happens in this Universe by itself. For every change, there is a cause. Every cause initiates a change and results in an effect, and the effect never precedes the cause. Thus, they discovered the Principle of Causality. They continued to study various phenomena, and started discovering physical laws in terms of the cause-effect relationships that govern the process of change. Scientists' faith in science strengthened over time as they discovered more laws, which not only explained, but also accurately predicted, the behavior of several phenomena.

Typically, a scientist would first set up some definite goals in advance. Scientific or logical thinking produces knowledge, which points and leads to methodical action. The causal connections thus perceived may or may not achieve the intended results. The scientist thus struggles continuously until he succeeds, gives up, or passes away. Fortunately, others step in and the effort goes on.

Science tries to find rational explanations for observed phenomena, including the origin the Universe. Science not only seeks to discover the rules, or physical laws, but also seeks to unify different phenomena, reducing the number of connections and laws. In other words, science has set the goal of rational unification and searches for a grand unified theory, which would explain everything in the Universe. It is a lofty goal indeed! Einstein devoted the latter part of his life to it, and many brilliant scientists continue to work toward this goal. But success has eluded them so far.

Science restricts its inquiry to the phenomena that are not subjective and can be measured and verified independently. Scientists thus focus their attention on the material world, and leave the study of human behavior, conscience, emotions and feelings, etc. to philosophy and psychology. With the help of mathematics, scientists attempt to explain all phenomena occurring in nature.

Of course, nature does not solve mathematical equations. Nature seems to follow the Universal Principle of Change, which we will discuss later in this book. We could thus define science as methodical thinking, which establishes causal connections. While establishing such causal connections and discovering physical laws, it attempts to reconstruct the origin of the Universe, its evolution to the present stage, and to predict its future.

We might also define science as a set of rules, or the physical laws of nature, which explains different phenomena in nature in terms of causal connections. We measure the power of a physical law by how effectively it organizes our thoughts and how successfully it enables us to use our simplest observations to make our most powerful predictions. If a law predicts something that does not match our observations, it simply means that we put in something in the law that nature did not follow.

- Road Map for Unfolding the Mysteries of Science
Scientists march on and boldly go where a mortal, with limited knowledge, is afraid to go. They unfold the various mysteries and

developments that brought us to our present state. Science has made phenomenal progress in understanding and unifying the laws of nature, especially during the last few centuries. Scientists have been studying motion on Earth and in the sky. They have studied Earth's gravitational attraction, electricity and thunder in the clouds, magnetism, as well as the light from the Sun and the stars, and the fireworks in space, extending to the farthest reaches of the Universe. Fig. 1.1 gives the road map, which shows some important scientists and various important milestones that we crossed on our journey through science

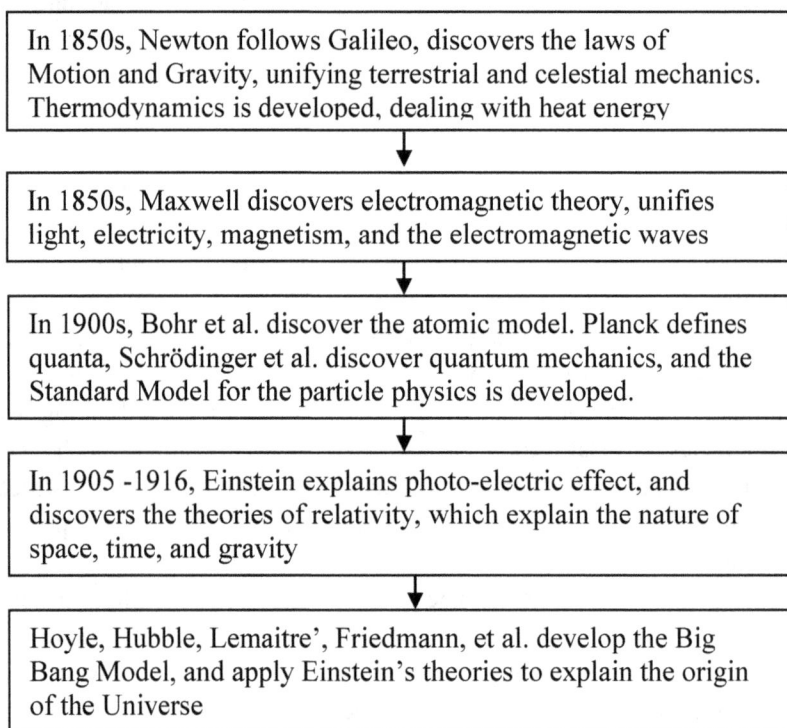

In 1850s, Newton follows Galileo, discovers the laws of Motion and Gravity, unifying terrestrial and celestial mechanics. Thermodynamics is developed, dealing with heat energy

↓

In 1850s, Maxwell discovers electromagnetic theory, unifies light, electricity, magnetism, and the electromagnetic waves

↓

In 1900s, Bohr et al. discover the atomic model. Planck defines quanta, Schrödinger et al. discover quantum mechanics, and the Standard Model for the particle physics is developed.

↓

In 1905 -1916, Einstein explains photo-electric effect, and discovers the theories of relativity, which explain the nature of space, time, and gravity

↓

Hoyle, Hubble, Lemaitre', Friedmann, et al. develop the Big Bang Model, and apply Einstein's theories to explain the origin of the Universe

---

Fig. 1.1 – Road Map for Unfolding the Mysteries of Science

The scientists have studied various ingredients that constitute the universe. These include space, time, energy, and matter (composed of elements, mixtures, and compounds like air and water) on this planet, which affect our daily life. The journey continues as scientists come, contribute, leave the scene, and others follow.

We have crossed major milestones in the field of science, traveling on the shoulders of several giants. According to a poll of scientists conducted by Physics World magazine, the top ten physicists in history are listed as follows:

Albert Einstein	Galilei Galileo
Isaac Newton	Richard Feynman
James Clerk Maxwell	Paul Dirac
Ernest Rutherford	Erwin Schrödinger
Werner Heisenberg	Neils Bohr

- Brief History of Science

Let us commence our journey with Galileo, the first true scientist, who dared to question the prevailing religious belief that the Sun rotates around the Earth. Galileo said:

"I do not feel obliged to believe that the same God who has endowed us with sense, reason, and intellect has intended us to forgo their use."

Newton followed Galileo, and formulated the laws of motion and gravity during the middle of the seventeenth century. Newton was modest when he said:

"To explain all nature is too difficult a task for any one man or even for any one age. Tis much better to do a little with certainty, and leave the rest for others that come after you, than to explain all things."

The enormous success of Newton's theory led Laplace to claim the doctrine of scientific determinism. It assumes the existence of deterministic laws that could predict what happens in the Universe, if we knew the initial state at any one time.

Several scientists contributed to our understanding of heat, energy, and work, as thermodynamics was developed. Thermodynamics provides answers about the relationships between energy, heat, and temperature. As we move on to thermodynamics, we come across the problem of the random motion of millions of molecules. Next, we move on to electricity, magnetism, and light – the forms of energy that make this Universe accessible and visible to us. We visit Maxwell, who unified electricity and magnetism, as he gave a unified theory of electricity, magnetism, and light in 1860. Even Maxwell thought, two hundred fifty years ago, that the demise of science was near.

Maxwell said,

"In a few years, all great physical constants will have been approximately estimated, and that the only occupation which will be left to men of science will be to carry these measurements to another place of decimals."

We keep our focus on the macroscopic world, and visit Einstein. Einstein came up with his special and general theory of relativity in the beginning of the twentieth century. His theories revised our notions of time and space, and explained the origin of gravity. One of the underlying assumptions is that energy cannot travel faster than the speed of light. These postulates destroyed the notion of absolute time and space.

Einstein's theories considered the relative motion between two observers moving relative to each other at a fixed speed (special relativity) or a fixed acceleration - change in speed (general relativity). Einstein believed in an orderly Universe and human fallibility: He said:

"The most incomprehensible thing about the Universe is that it is comprehensible."

"Anyone who has never made a mistake has never tried anything new."

"Only two things are infinite, the Universe and human stupidity, and I am not sure about the former."

Einstein explained gravity in terms of the curvature of space-time. When a body moves in space, it affects the curvature of space, which in turn affects how a body would move. In other words, time and space affect everything, and everything affects them in this Universe.

Then, we visit Rutherford and Bohr, who probed the microscopic world of matter and came up with an atomic model. Rutherford won the Nobel Prize in Chemistry in 1908, and Bohr received it in 1922.

We visit Planck, who came up with the notion of the quanta of energy, like coins, to account properly for blackbody radiation. He advanced the concept of packets of energy, called quanta. Planck regarded his hypothesis as a mathematical trick and not a physical reality. Planck won the Nobel Prize in Physics for this work in 1918. He said "Experience will prove whether this hypothesis is realized in nature."

As we visit modern physics, we find that the idea of quanta has gathered momentum and Planck's assertion has led us to quantum mechanics – a field that very few people understand. Schrödinger came up with Schrödinger quantum mechanics equations, ignoring gravitational effects for the elementary particles.

It has turned out to be a very successful theory for explaining the Standard Model for the atom and the microscopic behavior of the Universe. Schrödinger came up with his famous wave equation during a vacation trip to Switzerland with his girlfriend. In 1926, he published it and solved the standing wave problem for the hydrogen atom, leading to the birth of quantum mechanics.

We also meet other scientists who contributed to the development of quantum mechanics, such as Dirac and Feynman, etc. Dirac received the Nobel Prize for Physics in 1933 for his work on relativistic quantum mechanics. When asked about the beauty of a mathematical theory of physics, Paul Dirac replied,

"If the questioner was a mathematician then he did not need to be told, but were he not a mathematician then nothing would be able to convince him of it."

In the Introduction of the book, QED, Feynman presents an example, which shows the agreement to 12 significant figures between quantum theory and the prediction for the magnetic moment of the electron. Some famous quotes of Feynman:

"The test of all knowledge is experiment."

"One does not, by knowing all the physical laws as we know them today, immediately obtain an understanding of anything much."

"I love only nature, and I hate mathematicians."

Our journey takes us to Heisenberg, who introduced the Principle of Uncertainty. This Principle introduces an element of uncertainty, as quantum mechanics and quantum field theory explain the atomic and subatomic phenomena. We bring in probabilities to describe uncertainties observed at the microscopic level. The element of uncertainty introduced by quantum mechanics immediately raises questions about Laplace's doctrine of scientific determinism.

Laplace believed in a deterministic Universe. He said that if we knew the positions and speeds of all the particles in the Universe at one time, then we could calculate their behavior at any other time, in the past or future. Heisenberg agreed with this statement, but he said that one could never know the exact position and speed of a particle at a particular time. Heisenberg wrote, in his paper on the Uncertainty Principle,

"If we know the present exactly, we can calculate the future - it is not the conclusion that is wrong but the premise."

As science continues to move forward in the nineteenth century, many scientists think that the end of physics is imminent. Nobel Prize winner in physics, Max Born, said in 1928 that physics, as we knew it, would be dead in six months. However, more empirical data was soon collected that could not be explained with the help of the existing laws of physics. According to Born, quantum mechanics could not give exact results, but only probabilities for the occurrence of a variety of possible results.

"If God has made the world a perfect mechanism, He has at least conceded so much to our imperfect intellects that in order to predict little parts of it, we need not solve innumerable differential equations, but can use dice with fair success."

In modern physics, we spend some time visiting the Standard Model, based on quantum field theory – a combination of quantum mechanics, special relativity theory, and the quantization of the electromagnetic field. It attempts to describe all matter and fields, except gravity. Our journey takes us to the discussion of force fields and the efforts towards finding a unified field theory. We find that Einstein was quite unhappy about the notion of apparent randomness in nature. He said that God does not play dice with the world. Einstein once said to Heinrich Zangger in 1912 that the more success the quantum theory has, the sillier it looks.

Einstein felt that particles should have well-defined positions and speeds in reality, and should evolve according to deterministic laws. The uncertainty arises provisionally, because the quantum nature of light prevents us from seeing the reality. Ignoring quantum mechanics, Einstein attempted to develop a theory to unify both the theory of gravity and the theory of electromagnetic fields. Einstein believed that it should be possible to describe the Universe through one simple mathematical equation.

He devoted 35 years of his life to this quest and failed. A recent exhibition at the Natural History Museum displayed Einstein's last notebook. It contained his final calculations in pursuit of his Unified Theory, just before his death in 1955.

We then find that scientists discovered two additional fields with the help of the Standard Model, and called them strong and weak fields. Scientists then came up with the so-called Grand Unified Theory that combined the electromagnetic field, and the strong and weak fields. However, the search still goes on for a single theory to unify all the known fields, which would include gravity.

We also meet the present-day famous scientist, Stephan Hawking. He says.

"My goal is simple. It is complete understanding of the Universe, why it as it is, and why it exists at all."

"There is evidence that God is quite a gambler and the whole Universe is like a giant casino where dice are rolled on every occasion."

We find that the present goal of science appears to be the unification of two apparently contradictory theories, namely, the relativity theories explaining the behavior of large objects and quantum theory explaining the microscopic world. It becomes a necessity as scientists discover black holes, which need to be explained by quantum gravity theory - gravity theory due to extreme mass and quantum theory due to extremely small size. Efforts have started around developing quantum gravity theory. String theorists, believing in strings as the fundamental entity, claim some success, but scientists have yet to succeed in their final mission. Despite phenomenal progress, numerous problems remain unresolved and several questions are left unanswered.

The journey eventually leads to the Big Bang model for the origin of the Universe. We get an idea about the age and vast size of our Universe. Here, we visit the creation of matter, stars, galaxies, and black holes. Black holes fascinate us. On one hand, they are destructive monsters feeding on stars. On the other hand, according to some scientists, they might be responsible for the creation of stars and galaxies. We observe that Einstein's general relativity theory predicted the existence of black holes. We discuss how scientists recently confirmed their existence at the center of each galaxy. Then, we ponder over what happened before the Big Bang, and discuss some problems with the Big Bang model. We visit inflation theory, six numbers that decide the fate of the Universe, and the possibility of the existence of more than one Universe.

We then move on to list the remaining unanswered questions and the top ten mysteries of science: The questions regarding Newton's classical mechanics and gravity, concerning the true nature of inertial and gravitational mass, causality, and the mechanisms that transfer energy in 'free' space from one place to another. The unanswered questions concern the electromagnetic field, the true nature of photon, electric charge, quanta of energy, and the mechanism of electromagnetic wave propagation in free space. We also look into unanswered questions in modern physics, concerning the origin of various particles including field carrier particles, wave-particle duality, and the Uncertainty Principle, as we introduce probability into science through quantum mechanics.

In relativity, the remaining questions concern the constancy of the speed of light, the equivalence of inertial and gravitational mass, and the incompatibility of quantum mechanics and gravity. In cosmology, the unanswered questions concern the possibility of additional particles, dark matter, dark energy, the beginning and the end of the Universe, and the true nature of space, time, reality, and black holes. Towards the end of our journey, we summarize various milestones we crossed during our journey. We visit the frontiers of science and discuss its future. Finally, we attempt to discover the ultimate truth about the Universe.

1.1   *Basic Ingredients of the Universe*

Let's get ready for our journey by clarifying a few concepts, which include simple definitions of terms, such as energy, force, matter, space, time, information, and change. Scientists coin and define these terms to describe various phenomena observed in nature. Before we embark on this exciting journey, we must agree on the meaning of certain terms that are normally used to describe various phenomena. Various phenomena keep on happening, whether we have the terms to describe them or not.

We have coined terms for the basic ingredients of the Universe: space, time, matter, energy, and information. We also need to understand another important term, change. Change is eternal, perpetual, and it is the most important ingredient of every dynamic phenomenon. It involves time and the interplay of the remaining basic ingredients. Take any ingredient out, and the Universe ceases to exist.

. Perhaps, the first phenomenon that led early scientists to coin terms like space and time was motion – the change in position of an object around us as the day turned into night. How boring life would be if nothing changed! A scientist thus observes change with respect to time and/or space, and postulates that change requires energy. Thus, change occurs due to the play of energy in time and space. Scientists formulate theories and laws that govern change, and explain each phenomenon involving energy and force fields, matter, space and time.

How do we define these terms: energy, force, matter, time, space, information and change? For any entity, one can come up with various definitions - functional, operational, causal, mechanistic, and so on. Not only can we come up with different definitions, but also, every definition usually contains terms that need further definition. For example, a science teacher might define 'matter' as something that occupies space and has mass. A smart student would immediately ask for the definition of the terms, 'space' and 'mass'. The same sort of thing happens for almost every definition. We never resolve anything, as we keep defining one thing in terms of the other.

- Matter

As stated, the usual definition for matter is something that occupies space and has mass. This definition contains the terms 'space' and 'mass', which require definition. One might also define matter as a composite of different molecules - or atoms, or elementary particles, or just electrons and quarks (bound by gluons) - which are assigned certain properties, such as mass, charge, etc. One could further define each elementary particle as a unique singular energy configuration. One might say that matter, at the fundamental level, is the concentrated form of energy, or a composite of singular energy configurations.

We have to introduce the terms mass, space, energy, and elementary particles just to define matter. Then, we have to define 'dark matter'. We do not know much about it. It constitutes 23% of the contents in the Universe. According to some scientists, dark matter is some kind of stagnant gas. One needs it to provide the additional gravitational force to keep the galaxies from flying apart. The remaining 4% constitutes the mass of all the ordinary visible matter and radiation (a tiny fraction - 0.005%) in this Universe. Our planet Earth, like a speck of dust in this vast Universe, is an infinitesimal fraction of this 4% ordinary matter.

We know that one can convert ordinary matter into energy and vice versa. In fact, Einstein's famous equation, $E = mc2$, relates energy (E) to the mass (m) through the speed of light (c). We play with and control only part of the energy associated with the tiny fraction (4%) of ordinary matter. We transform it to get energy for our bodies, to drive cars, to generate electricity and nuclear power, and for heating or air conditioning our homes, etc. Of course, we never really spend this energy, since it only transforms into different forms of energy or matter or vice-versa. How do we define energy?

- Energy and Force

Let us define energy as an agent of change that is responsible for every phenomenon in this Universe. Energy is the prime cause of any change in this Universe. We need energy to accomplish certain tasks, such as moving objects, heating water, turning on the lights, or powering a television set. The usual scientific definition for 'energy' is the capability of an object or a system to do work on another object or a system. However, what is 'work'? The first scientific definition of 'work' in mechanics was in terms of moving material objects.

The concept of energy, or work (W), is closely linked to force (F) and distance (D) in space. Moving a material object in space over a distance (D) requires the application of force (F). 'Work' equals 'force' times 'distance', i.e. $W = F \times D$. Note that this definition of energy involves force, matter, space and time, all of which need to be defined.

What is 'force'? The usual definition of mechanical force (F) is given in terms of mass (m) and acceleration (a), $F = ma$. In fact, the concept of force is perhaps more fundamental than energy. Energy, believed to be stored in force fields, plays its role through the force fields. For example, the force exerted on our feet by the floor, is due to the electromagnetic force, which resists the displacement of atoms from the equilibrium position in matter. Scientists have discovered four force fields in our Universe, namely, gravity, electromagnetic, weak, and strong fields. We shall visit them during our journey.

Scientists believe that each of these force fields interacts through the exchange of their carrier particles. These force fields have energy associated with them, since they have the capacity to do work. We can also define the intensity of these force fields. For example, for an electrical force field, we can define its intensity at distance (r) in terms of the force required to move a unit charge at distance (r) in this field[1].

One can also define energy as a condition or a state of a physical entity. However, such a definition does not tell us much, unless we define the term 'state'. Other definitions tell us what the energy can do, but not what it is. Energy can propagate in space from one point to another point at a speed that doesn't exceed the speed of light. One can determine the energy carried by an electromagnetic wave. The units of measurement for energy are very confusing, as one can measure it in so many different ways. One of the famous scientists, Feynman, once said,

"For those who want some proof that physicists are human, the proof is in the idiocy of all the different units, which they use for measuring energy."

We know certain things about energy. It occurs in many forms. Regarding different forms of energy, most of these forms reside in the four fundamental force fields. We cannot create or destroy it, but we can store and transform it from one form to another, into matter and vice-versa. We have seen devices, called transducers, which convert or transform one form of energy into another. In fact, our body has built-in biological transducers that transform almost every form of energy into another form that's needed by our body. We have also developed different types of transducers that emulate those found in nature.

We use a series of transducers to convert one form of energy into another. For example, in telephony, we go from speech (sound) to pressure waves in the air, then to mechanical vibrations using a transducer (telephone transmitter) where a diaphragm converts pressure waves into vibrations. One converts these mechanical vibrations into electrical signals using a transducer (strain gauge or carbon granules inside the telephone transmitter). These transducers alter the resistance (and electrical signals) in an electric circuit.

Thus, one can transform speech into fluctuations of electrical signals. Next, we modulate these signals with a carrier, transmit them, and receive them on the other end. We go through the reverse process to recreate speech on the telephone receiver. Microphones are examples of such transducers, which convert sound to electrical signals, and loudspeakers convert electrical signals back to sound.

Table 1.1 lists some of the energy transformation techniques and the well-known transducers. Not included directly in Table 1.1 are some important forms of energy. These include gravity, vacuum energy, dark energy, and nuclear energy. Gravitational energy, although a small amount in the Universe, is extremely important. The motion of stellar objects, stars, and galaxies - revolving in their orbits - mostly involves gravity. We make use of gravitational energy in hydraulic power stations. In such power systems, we convert the potential energy of the water due to gravity, stored at a certain height, into kinetic energy and then into electrical energy. We do not yet understand vacuum energy or dark energy. Regarding nuclear energy, we first misused nuclear energy by making thermonuclear bombs. We then harnessed nuclear energy and built nuclear power stations, based on nuclear fission in atoms and converting the released nuclear energy into electrical energy.

Our Universe contains a vast amount of energy. We do not know the ultimate source of this energy, but it was present when the Universe originated with a Big Bang. The energy of the Universe comes in the forms of dark energy, radiation, ordinary matter, and dark matter. Scientists on the NASA WMAP team found, in early 2005, that the contents of the Universe include 4% ordinary matter and radiation, 23% dark matter, and 73% of a mysterious dark energy.

Scientists are quite perplexed about this dark energy. We do not know much about the 73% dark energy, except that it is behaving like anti-gravity. Gravity keeps the planets, stars, and galaxies together in orbits by attracting them towards each other. The dark energy tugs on the fabric of time and space, and pushes galaxies apart ever faster and faster into the farthest reaches of the Universe. The discovery of dark energy is revolutionary, as it confirms that the Universe is not merely expanding, as Hubble showed in the 1920s, but its expansion is accelerating.

ENERGY	Mechanical	Sound	Electric	Magnetic	Heat	Light	Chemical
Mech.	X	Drums	Electric Generator	Magnetize Iron	Friction Devices	Friction Devices	
Sound	Telephone Transmitter	X	Micro-phone				
Electric	Electric Motor	Loud-speaker	X	Electro-Magnet	Electric Heater	Light Bulb	Electrolysis
Magnetic	Magnetic-levitation Train			X			
Heat	Steam Turbine		Heat Two Metals Junctions		X		
Light	Laser Gun		Solar Cell		Solar Heater	X	Plants
Chemical	Heat Engine		Batteries		Gas Heater	Phospho-r-escence.	X

Table 1.1 – Energy transformation and transducers

It is interesting how we can denote the energy that's given off by an object or system as 'positive', and energy gained by an object or system as 'negative'. Scientists, for example, denote gravity as negative energy and matter, or dark energy, as positive. Since we cannot create or destroy energy, different forms of energy always maintain the balance throughout the process of transformation. That is why some scientists claim that the total energy in the Universe is constant, and the net algebraic sum of the total energy of the Universe is zero!

- Space and Time

How do we define space and time? Is space a container of energy and matter? Is it a physical entity, filled with some kind of 'ether', or filled with quantum energy and virtual particles? Or is it empty, devoid of anything - a true vacuum? It is ironic that scientists are still struggling to understand the true nature of space and time, and they are still trying to come up with clear definitions.

For example, there are two differing views regarding space and time. One group claims that space and time are substantial entities, and they exist independently of matter and energy. The second group contends that they are merely artificial means to describe how objects are related.

One definition, usually given for time, is that time is a continuum in which one event follows another from the past to the future. However, what are 'past', 'present', and 'future'? According to Einstein, they are mere illusions. His theory of relativity says that even simultaneity is relative. In other words, observers moving relative to each other associate different meanings with the present, past and future. The distance contracts and time slows down for one observer moving relative to the other observer. The distance and time have to change for observers moving relative to each other, if the speed of light is to remain constant in space.

We shall address these issues soon, and revisit them often throughout the book. For now, let us just say that time has a definite meaning only when energy, or a particle, moves in space with a certain delay. We measure this delay by comparing it to another event, and call it 'time'. Energy and matter perform their act on the stage of space and time.

- Information

In scientific circles, scientists realize that information is also an essential ingredient in the Universe. If there is little change, there is little information. A single frequency tone has very little information, compared to a voice signal. How do we define 'information'? Simply stated, it is the minimum data bits (0 or 1) that are needed to characterize something completely.

Shannon, in 1948, in a paper in Bell Technical Journal, unified all phenomena related to classical information processing by quantifying the information content produced by an information source. He defined it to be the minimum number of bits needed to store the output of the source reliably. His expression, known as Shannon entropy, plays a central role in data compression, information transmission over noisy channels, and in analyzing stock market behavior, or any system behavior that processes information.

We shall return to information science while discussing the evolution of our Universe, and genetic and cultural evolution, in the third chapter. We will also revisit the field of information science during our journey through technology in my second book, especially, communication and computer technology, when we explore the possibility of quantum computing.

- Change

Finally, we define the term 'change' that is responsible for every dynamic phenomenon in the Universe. Everybody in this Universe is always on the move, and changing. The Universe that we observe today is the result of continuous changes in the states of various entities. Since the Universe is always changing, the most important mission of science is to formulate theories and to discover the laws that govern the process of change in terms of cause and effect. In the frontiers of science in the fourth chapter, we shall propose a Universal Principle of Change, which is applicable to both inanimate and animate objects.

What does the term 'change' mean? We associate a state with everything in this Universe, e.g., coordinates with a point in space, temperature with an object, and charge and mass with an electron, etc. Change occurs when the state assumes a value different from its present state. This change can be in position, temperature, or in a state of any entity. We always define change with respect to, or relative to, something. The process of change generally involves energy (force field), matter, space, and time. Usually, it involves time and space, since one observes change over a certain period in a certain region of space. In fact, change makes us aware of the flow or passage of time.

For example, if we observe an object and its position remains the same with respect to time, we say that there is no change in position. On the other hand, if the position of the object changes uniformly with respect to time, we say that the position is changing with respect to time at a constant rate. However, another state of the object, namely, its speed (rate of change with respect to time) is not changing. One might also say that, as far as the position of the object is concerned, we do not have an equilibrium state, but its speed has reached an equilibrium state. Suppose the position (x) was changing periodically at a frequency (f) in a sine wave fashion, as in $x(t) = x_o.\sin 2\pi ft$, with respect to time (t). Then, we say that its position (and speed) is changing with time, but in a steady, sinusoidal manner.

Why does change occur? Energy is responsible for every change. Change occurs when we transform one form of energy into another. The interaction of force fields and the exchange, or transformation of energy, causes change. It actually takes place when two or more force fields interact with each other in a region of space. As these force fields start interacting, the process of energy transformation begins in that region of space, which initiates the process of change at once. This force field interaction, resulting in the energy transformation, is thus the primary cause of change. During the process of change, the state of an object (e.g. position) in that region of space keeps on changing, until the process of energy transformation is completed and equilibrium is established.

Now that we have gained some understanding of certain important terms, we are ready to start our journey. We shall return and discuss the fundamental nature of space, time, energy, matter, and change towards the end of our journey, while visiting the frontiers of science.

1.2   Motion, Gravity and Newton

- Motion

One of the first things we notice on Earth is motion – a child observes motion as we move around him. When a child throws a ball up in the air, it falls back to Earth. It is one of the most fundamental characteristics of the Universe. We observe motion in the Universe as moons, planets, stars and galaxies revolve and move in different orbits. It would be unbearable if we were just a fixture and could not move in space, and change our position in space with time.

Aristotle studied the motion of the Earth, Moon, and Sun as early as 340 BC. He believed that the Earth was stationary, and the Moon and the Sun revolved around it. Galileo and several other scientists, including Copernicus and Kepler, observed and studied the motion of the Moon, planets, and stars. They established that the Earth revolves around the Sun. We now know that every entity is in a state of motion relative to something else, and nothing in this dynamic Universe remains stationary. Earth is in motion, and our entire planet rotates on its axis and revolves around the Sun.

While observing motion, we notice that a moving object keeps moving, unless stopped by some force. We also notice that the more force we exert to push an object, the faster it moves. Furthermore, when we push an object away, our hand feels the pressure exerted by the object. Newton formalized these observations in the form of the three laws of motion. His book, Principia Mathematica, published in 1687, is one of the most influential books in physics. Newton also formulated gravitational law, and discovered and developed calculus, which is important to studying change in any quantity with respect to time and space.

Simply stated, Newton's laws of motion follow from the simple principles that there can be no effect (or change) without a cause (or expenditure of energy), every cause has a definite effect, and every effect causes an equal and opposite reaction. We can formally write Newton's laws of motion as follows:

1) *An object continues in its state of rest or motion with uniform velocity unless an unbalanced or net external force acts on it.*

2) *The net force acting on an object of mass accelerates it to a value given by the relation: Force = mass x acceleration.*
3) *Forces always occur in pairs. If object A exerts a force on object B, then B exerts an equal but opposite force on A.*

Newton developed his classical mechanics theory based on these three laws of motion, which explains all motion. The theory of classical mechanics addresses the motion of matter when subjected to mechanical force. The theory connects mass and force to kinetic quantities - displacement, velocity, and acceleration. However, it does not address the true nature of mass, force, space, or time. We shall discuss this subject when we discuss the frontiers of science.

The laws of motion, based on observations, have been very successful in explaining issues related to material objects moving at speeds much lower than the speed of light (c). Newtonian mechanics is non-relativistic in that the moving object and the observer are in the same frame of reference. For objects traveling close to speeds approaching the speed of light relative to the observer in different reference frames, Einstein provided the correct answers in his special and general theories of relativity.

- Gravity

We experience gravity as we see objects falling towards the ground. We have all observed objects falling to the ground, but it was Newton who linked the falling of an apple to the force of gravity, and came up with the law of gravitation. Interesting enough, Aristotle believed that heavier objects fall faster than lighter objects towards the Earth. Galileo proved him wrong when he took the trouble of performing a simple experiment, dropping a heavier and a lighter body simultaneously from the Tower of Pisa. They both landed on the ground almost at the same time. The difference in travel time was only due to the difference in air resistance for different-sized objects.

The discovery of gravity by Newton is an interesting story. According to John Conduitt, Newton, while musing in a garden in 1666, after retiring from Cambridge, saw an apple falling from a tree. A thought entered his mind that perhaps the same force pulled all falling objects downward. Newton named this force 'gravity'. He then realized that the power of gravity, which brought an apple from a tree down to the ground, was not limited to a certain distance from Earth, but extended much farther than was usually thought. It was the first giant step towards unification in science. Newton then thought about the Earth orbiting around the Sun by considering circular motion.

The first law of motion implies that, in the absence of any forces, motion in a straight line with constant velocity continues indefinitely. However, if we tie a stone to a string and hurl it around, it moves around at velocity (v) in a circle of radius (R), as shown in Fig.1.2.

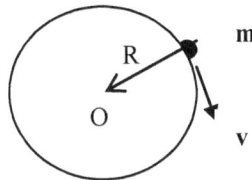

Fig. 1.2 – Stone on a string in circular motion

Motion in a circle is different. If an object moves around a circle, we need a force to maintain that motion—otherwise, it flies off at a tangent, with constant velocity, along a straight line. One calls the acceleration (a) towards the center, needed to keep an object moving in a circle, centripetal acceleration. It can be determined from Newton's second law of motion ($a = v^2/R$). By Newton's laws, any acceleration requires a force (F) that must constantly pull the stone with inertial mass (m) towards the center.

This force ($F = ma = m \, v^2/R$) is called the centripetal force. To keep the stone going in a circle, we must provide this centripetal acceleration through a force directed towards the center. We do so through the string, by pulling it towards us and keeping it stretched in tension. The force keeps the string stretched by continually pulling the stone towards the center. If the string breaks, the stone would fly off with velocity (v) in a straight line along the tangent to the circle.

Newton observed the Moon orbiting around the Earth. Perhaps, he reasoned that, since the size of Moon does not appear to change, its distance stays about the same, and hence, its orbit must be close to a circle. The Earth must exert a pull on the Moon to keep it moving in that circle. This centripetal acceleration towards the center of the Earth is necessary to keep the Moon in a circular orbit. Newton must have reasoned that the force of gravity provides this acceleration. In fact, the force makes the apple fall to the ground. In other words, the Moon keeps gravity engaged in order to keep its circular orbit.

Newton showed that gravity at a distance (R) was proportional to $1/R^2$ (varied like the 'inverse square of the distance'). The acceleration (g) measured at the Earth's surface would correctly predict the orbital period (T) of the Moon. Thus confirming the 'inverse square law', he proposed that a 'universal' attractive force of gravitation (F) exists between any two masses, (m) and (M), directed from each to the other.

His law of gravitation essentially states that every object exerts an attractive force (F) on every other object. The magnitude of this force is directly proportional to the gravitational masses of two objects, (m) and (M), and inversely proportional to the square of the distance (r) between their center of gravity ($F = G.m.M/r^2$).

We call the constant of proportionality (G) the universal gravitational constant, which has a very small value, 6.6726×10^{-11} N.m/kg^2. Suppose (M) denotes the mass of Earth, and (r) is the distance of an object of mass (m) from the center of the Earth. Then, the gravitational force (F) attracting the object is given as $F = m g$. We call it the weight of the object. The g ($= G.M/r^2$) is acceleration due to Earth's gravity. Rotating a centrifuge, or during a space vehicle liftoff, we can also create very high values for the acceleration corresponding to several g's.

Our Universe contains a large number of massive objects (e.g. stars). Cosmologists learn a great deal using Newton's theory, which gives very good results as long as the patch of the Universe we want to study is small compared to the Universe horizon. For example, a simple application of Newton's theory explains the cosmic web structure in terms of gravitational instability phenomenon. We observe that a hierarchy of structures weaves this cosmic web, when we observe the Universe starting at a few million light years all the way to almost 14 billion light years away (age of the Universe ~ 13.7 billion yr.).

To explain this clustering, or web phenomenon, let us assume that we have three equidistant and equal masses, as shown in Fig. 1.3.

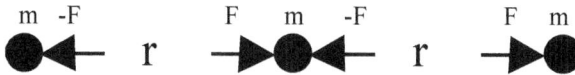

Fig. 1.3 - Gravitational instability

The force of attraction felt by each mass due to gravity is given as, $F = G.m^2/r^2$. The mass on the left and right attracts the mass in the center. Thus, it experiences zero net force due to gravity. However, if we perturb the center mass a little from its position, it would experience a net gravitational force. We can extend this logic to a very large number of objects in the Universe. The slight fluctuations in the mass density of such objects in space, leads to gravitational force field intensity. The resulting instability is responsible for the web structure.

Einstein observed that the gravitational mass of an object measured from the gravitational law equals the inertial mass of the object, measured from Newton's second law of motion. Einstein also defined gravity in terms of the curvature of space-time. We still do not know the exact mechanism through which the gravitational field interacts. Scientists suspect that it happens through the exchange of particles called gravitons, but no one has detected a graviton yet. We shall be revisiting gravity often during our journey.

1.3   Heat and Thermodynamics

We are well aware of the heat from the Sun, which is our source of thermal energy. Thermal energy interacts with the environment, which changes the weather and the dynamics of the Universe. Thermodynamics deals with thermal issues. It is the study of the relationship between energy, heat, and temperature. It also tells us how the transformation between these quantities takes place, and what the role is of entropy (disorder) in physical processes. In fact, the invention of engines spurred the development of thermodynamics in the 18th and 19th centuries. Thermodynamics gives us the notion of temperature – the measure of the degree of heat, calories to measure the quantity of heat, and the relationship of heat to work. It also gives us the notion of entropy, disorder, or loss of information, and the understanding of gas dynamics and heat engines, etc.

The laws of thermodynamics are very important. These laws deserve our attention during this journey. The first law is essentially a statement about the conservation of energy. Applied to thermodynamics, it states that the net heat added to a system equals the change in internal energy of the system and the work done by the system. In other words, when we heat a system, it raises the internal energy of the system and converts the remaining energy to work.

The second law essentially states that entropy, or disorder, must always increase in any closed system. Thermodynamic equilibrium is the state of maximum disorder. The loss of information accompanies the increase in entropy or disorder. A very disordered state needs only a few bits of information to describe it. For example, the macroscopic state of a gas in thermodynamic equilibrium requires only temperature and volume for its description. However, gas that's not in a thermodynamic equilibrium has hot spots and eddies, and requires more information for its description.

Scientists have realized, around 1858, that the second law of thermodynamics is nature's way of driving a closed system to equilibrium, maximum disorder, or minimum information. A perfume, or gas in a bottle (ordered state), once released to the outside air, can never go back into the bottle on its own, as it mixes with the outside air, involving innumerable intermolecular collisions, causing increasing disorder. Applied to heat, the second law implies that, in an isolated system, heat always flows from hot to cold and the result is thermodynamic equilibrium, with heat evenly distributed at a uniform temperature.

The second law raises an intriguing question. Why should every closed system follow this law? Why should it be stuck with a one-way ticket on a slide towards complete disorder or chaos? Scientists say - simply because there are many more disordered states than the ordered states. For example, a new deck of cards that's neatly ordered, when shuffled a little, very easily becomes highly disordered as far as suits and the numerical sequencing are concerned. In other words, disturbing an ordered state is most likely to produce less-ordered states and not vice-versa. This fact is the source of the directionality towards disorder in an irreversible or time-asymmetric manner, although all the laws of physics are time-reversible or time-symmetric.

The second law of thermodynamics applies to many fields, including black holes and genetics, discussed later. For example, in genetics, it applies to the loss of information in cells every time they reproduce, or when external chemical agents affect them. As entropy increases and information is lost, cells do not have vitality. Eventually, it results in aging, as cells are damaged and they cannot repair themselves. Our bodies and nature's creations - as well as stars, black holes, and the Universe - must obey this law and reach a stage of maximum disorder or chaos. New ideas continue to be developed, extending its frontiers. For example, a book by Ralph Tykodi, Thermodynamics of Systems in Non-equilibrium States, discusses some very interesting new ideas.

1.4 Light, Electricity and Maxwell

Light does not need a physical medium, and it can travel in free space. Sound, on the other hand, does need a medium, since sound waves are compression waves and travel by compressing the elements of the medium. Without light, the Universe would be dark and invisible. Without sound, there would be dead silence in the Universe. Without electricity and the associated magnetic phenomenon, we would not have modern means of communication - no TV, no phones, no cars, etc.

We have often wondered about the lightning from the clouds, video and audio electrical signals plucked from the air, and the light from the Sun or an electric bulb. It is difficult to imagine life without a simple electric lamp or a basic telephone. Even our Earth is a magnet, and a compass needle always points towards the magnetic North Pole. Electromagnetic radiation holds the key to the functioning of our Universe. In fact, without this field, life would simply cease to exist.

Learning and Communication
To observe or learn about any object, we need three things:
1. Source(s) or transmitter(s)- source of energy or information
2. Object or target - object to be studied
3. Detector(s) – to collect the wave reflected from the object

For example, a wave travels from the source (e.g. light source)), bounces off the object (e.g. box), and the reflected wave travels back to the receiver or detector (e.g. eye retina). The information is registered inside the receiver, and converted to the appropriate signals (signals from the retina carried to the brain) for processing by the equipment (brain). The processor (brain) analyzes the information and projects a visual image of the object (e.g. box) in our mind. Thus, our mental model of the object helps us perceive the reality around us.

A person can see things because objects scatter the light that strikes them, reflecting some of it back to the eye. If one could reduce an object's reflection and its shadow, we could have a cloaking device that would make the object invisible. One could conceptually design cloaking devices for any specific bandwidth of radiation. In fact, Schurig and Smith at Duke University successfully cloaked, in 2006, a copper cylinder from microwaves. They used meta-materials that channel the microwaves around the cylinder.

Waves

We thus perceive the reality of the physical world through waves, because the Universe is awash with all kinds of waves.

Let us first clarify a few basic notions about waves. A wave wiggles in space-time and travels through space. Propagating waves carry energy and information from one place to another. Life would simply cease to exist without waves, as we could not receive energy from the Sun.

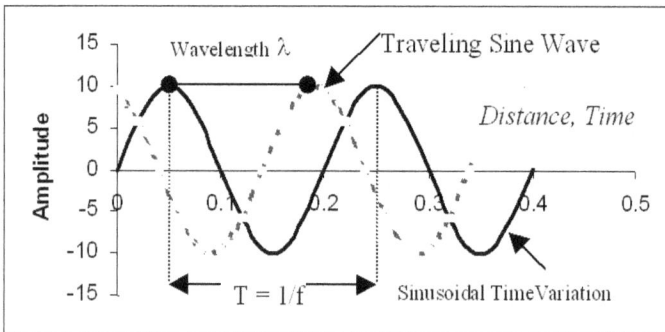

Fig. 1.4 – Wave motion (sine wave): Asin $(2\pi x/\lambda - 2\pi ft)$

Fig. 1.4 shows a sine wave, $A_i \sin(2\pi x/\lambda - 2\pi ft)$. Maximum amplitude A=10; Wavelength λ = 20 cm; Frequency f = 5 Hz. It is traveling in space in x-direction at a speed of v (= fλ) 100 cm/sec, while varying in a sinusoidal manner in time.

The sine wave is the most fundamental wave in this Universe, since we can express all other waves in terms of sine waves of different amplitude, frequency, and phase. A wave has the following properties:
1) The maximum field fluctuation is the amplitude (A).
2) The number of field changes in one second is frequency (f).
3) The distance wave travels between two crests is the wavelength (λ).

A wave propagates at a set velocity, v = (f. λ), through a material. We can always express a periodic (repetitive) wave as the summation of sine and cosine waves of different amplitudes with multiple frequencies, called harmonics, with the help of the Fourier series. We can express any non-periodic (non-repetitive) wave in terms of a continuous frequency spectrum using Fourier transform.

Two waves can pass through each other and cause 'interference'. Fig. 1.5 illustrates the phenomenon of interference. Two waves in phase cause constructive interference. The same waves, when out of phase, cause destructive interference.

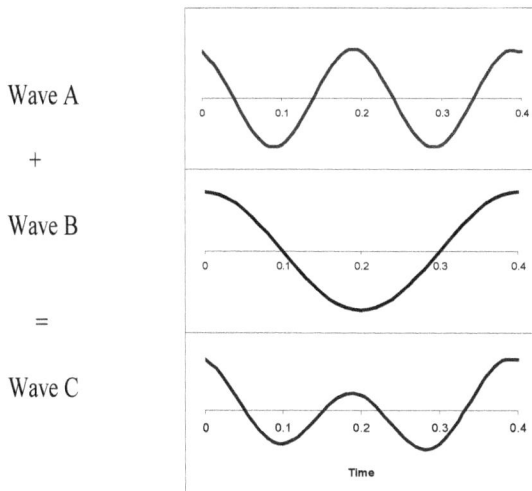

Fig. 1.5 – Wave C: Result of interference of waves A & B

Modern-day wireless communication owes its origin to the propagation of electromagnetic waves. We convert information into an electrical signal, using a transducer (e.g. microphone). We then modulate it with the carrier frequency and transmit it, riding on carrier waves traveling at the speed of light, in one of the different frequency bands allocated for transmission. On the receiving end, demodulators and other devices inside a receiver extract the information from the received signal.

Governments allocate and sell the frequency bands to communication companies, as shown in Fig. 1.6. The frequency spectrum has expanded all the way to the optical range. Nevertheless, as it expands, it gets crowded, as the demand for communication grows.

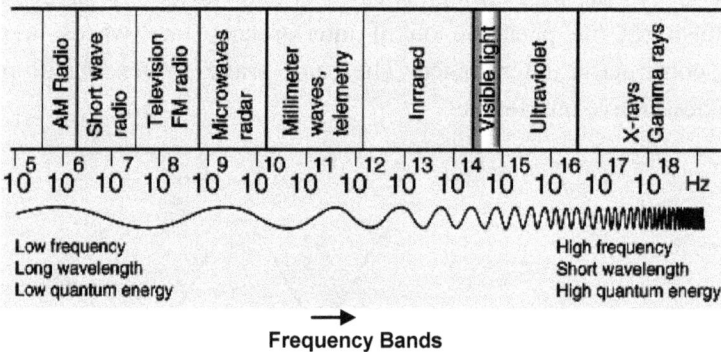

Fig. 1.6– The Electromagnetic Waves Spectrum

Several scientists have contributed to our understanding of electricity as well as magnetism. The contributors include names like Faraday, Franklin, Ampere, and even Newton for the theory of optics. However, it was Maxwell who unified electricity and magnetism around 1860. Maxwell postulated that electricity and magnetism intertwine. Wherever there is electricity, there will be an associated magnetic field and vice-versa. He showed that light itself is an electromagnetic phenomenon, and electromagnetic waves travel in free space with the speed of light.

Now, scientists say that light energy consists of trillions of photons, quanta of energy, but these photons have wavelike properties.

Maxwell formulated the theory of electromagnetism and put it in a concise mathematical form, now known as Maxwell's equations. These equations express the physical laws for the electromagnetic phenomenon. They give a complete description of electromagnetic phenomena, and predict the propagation of electromagnetic waves. One can derive the electromagnetic wave equation from Maxwell's equations.

The Maxwell's first equation simply expresses the relationship between an electric charge and an electric field. The second equation expresses the absence of isolated magnetic charge, and the third and fourth equations express the fact that time change in a magnetic field causes an electric field and vice-versa. For mathematically inclined readers, these equations are written at the end of the book[2]. Readers without a mathematical background can also enjoy their beauty and elegance.

How do electromagnetic fields interact? Quantum theory provides an intuitive picture of an electromagnetic interaction as an exchange of photons between two interacting objects. For example, when we bring a magnet close to another, they exchange photons. This exchange leads to the force between them. An electric (magnetic) field produces an attractive/ repulsive force between charged objects (magnetic poles), which vary as the inverse square of the distance between them.

In fact, different atoms interact together to form molecules and chemical compounds because of the residual electromagnetic force, due to the charged particles in each atom. All the interactions that are needed to explain chemistry and biology are electromagnetic in nature. In other words, the electromagnetic force is responsible for all chemistry, biology, and life!

Do electromagnetic waves require a medium for propagation? Maxwell thought that electromagnetic waves, just like sound waves, needed a medium for propagation. In the 1878 edition of the Encyclopedia Britannica, he described ether as a medium for the propagation of electromagnetic waves. In this article, he also recalled a failed attempt to measure the effect of the ether drag on the Earth's motion. He proposed an astronomical determination of the ether drag by measuring the velocity of light using Jupiter's moons at different positions relative to the Earth. Inspired by Maxwell's ideas,

Michelson began his own experiments. He reported in 1881 that the hypothesis of stationary ether is incorrect and erroneous. Michelson and Morley refined and repeated their experiment many times until 1929, but came up with the same conclusion. These experiments showed that the velocity of light was independent of the velocity of the observer. This raised questions about the measurement of time and distance.

It led Poincaré to suggest that the simultaneity of two events, or the order of their succession, as well as the equality of two time intervals, must be defined so as to keep the laws of nature simple. Poincaré once said,

"Science is built upon facts, as a house is built of stones; but an accumulation of facts is no more a science than a heap of stones is a house."

Poincaré's observations on Michelson and Morley's results and asymmetry, observed in the electromagnetic phenomenon, finally led Einstein to propose the theory of relativity, changing the notion of time and space.

1.5   Relativity, Gravity and Einstein

Towards the end of the nineteenth century, Einstein, a 26-year-old young clerk, with his doctoral thesis rejected, started working in a patent office in Germany. In his spare time, he used to think about space, time, and motion. He was uncomfortable with the Maxwell equations. Specifically, he did not like the discrepancy in the explained behavior of a magnet moving relative to a conductor and the conductor moving relative to a magnet. He felt that all motion is relative, and the laws of nature should be symmetric with respect to space and time. Thus, we must be able to explain relative motion between a magnet and a conductor in an identical manner.

Einstein was a strong believer in the elegance, beauty, and symmetry of nature. He argued that, if two observers are moving relative to each other, then nature should not treat them on unequal footings. In other words, all physical laws should be the same, regardless of one's state of motion, as suggested by Galileo back in 1632. In the 1890s, Lorentz also studied the invariance of physical laws for different observers, representing the symmetry of space and time - called Lorentz symmetry. Lorentz symmetry forms the core of relativity theories. It is also the foundation of the Standard Model of particle physics. We obtain the Standard Model when we combine Lorentz symmetry with quantum mechanics.

Einstein thought about the observation that the speed of light in free space does not change, no matter how fast or in which direction an observer moved while measuring it. He came up with the relativity theories when he considered observers moving relative to each other, and observing time and space in different reference frames. Since speed equals distance divided by time, and the speed of light cannot change, both distance and time cannot remain the same. Both must change for the observers watching each other in two different reference frames, while moving at a uniform speed relative to each other.

By some miracle of nature, the distance or length shrinks and the time or the clock slows down to keep the speed of light constant. Einstein's theory of special relativity thus demolished the commonly accepted notions of space-time as unchanging and absolute. In short, the basis of the theory of special relativity is essentially the symmetry of the laws of nature with respect to uniform relative motion and the constancy of the speed of light in free space.

In 1907, Einstein also had the happiest thought of his life, when he realized that gravity and acceleration are indistinguishable in a local frame of reference. He went on to develop the theory of general relativity, which redefines gravity in terms of the curvature of space. It changes the commonly accepted notion of gravity advanced by Newton.

Two simple ideas - namely, that gravitational mass is the same as the inertial mass and all the physical laws are the same in all reference frames - combined with the theory of special relativity, led to the theory of general relativity. To sum up, the relativity theories discuss the relative motion of two observers in different reference frames. Uniform speed leads to the special theory of relativity, and uniform acceleration leads to the general theory of relativity.

The special relativity theory radically changes our views of space and time. The distance and the time, measured by two observers moving relative to each other, are not the same. Time and space are warped; the faster an observer moves, the slower his clock ticks, and the shorter his ruler becomes to measure the distance. The theory of general relativity explains gravity in terms of the curvature of the space-time fabric. It predicted that the gravitational field also propagates at the speed of light and not instantaneously, as Newton thought. The general theory of relativity also led to the equations that explained the origin of the Universe in terms of Big Bang model.

- Theory of Special Relativity

Although the mathematical details of the Einstein theories are not easy to understand, the basic ideas are simple. While traveling in a train, or in an airplane flying at uniform speed, we move inside and behave as if the train or plane is not moving. We are stationary with respect to the moving object because we are part of it. Similarly, we are stationary with respect to Earth, although it revolves around the Sun. It also rotates on its axis, completing one rotation every 24 hours. A point on the Equator moves eastward at a speed of over 1600 km. every hour. We do not feel this motion since we are not moving relative to Earth.

The problems arise only when there is relative motion. For example, if we were sitting in a stationary train and watching another train pass by, we might get the feeling that our train is moving. The person traveling in the other train would have the same feeling, since motion is relative. Einstein studied such relative motion, involving two observers in different reference frames moving relative to each other. He made startling discoveries. The observers moving relative to each other might even disagree on their measurement of distance and time. This disagreement on time and distance becomes quite large as their relative speed approaches the speed of light. Einstein showed that this disagreement is due to the very nature of space and time, dictated by the constant speed of light.

We can state the basic postulates of the special theory of relativity for objects moving at uniform speed relative to each other, as follows:

In every inertial frame of reference:

1. *The physical laws of nature are the same*
2. *The speed of light is the same*

These postulates essentially use Lorentz symmetry to describe the laws of physics in our Universe. Lorentz symmetry, discussed later, is essentially the symmetry that requires that the laws of physics be the same for all observers.

We define the 'inertial frame of reference' as one that is at either rest or moving in a straight line at a constant speed. For example, if the airplane is moving at a constant speed in a straight line, then the laws governing the thrown objects inside are the same, as if the plane were at rest. We call the inside of the plane an 'inertial frame of reference'. It is also interesting to note that the constant value for the speed of light follows directly from Maxwell's equations, describing all phenomena related to electricity and magnetism, including light. It is a simple consequence of applying the first postulate to Maxwell's equations.

As stated above, one of the interesting consequences of special relativity is that an observer (A) moving toward another (B) at a uniform speed observes the relative motion affecting time and space. Each observer moving relative to the other observes that time slows and space shortens in the other observer's reference frame. In other words, observer (A) discovers that observer B's clock ticks slower than his clock, and the ruler that observer (B) carries is shorter than his ruler. Observer (B) observes exactly the same things about A. Thus, time and space measurements are not the same, when observed by observers moving uniformly relative to each other. Both are affected, and the effect becomes more prominent when the relative speed of the two observers approaches the speed of light.

- Theory of General Relativity

It is only natural to ask the next question: what happens when the two observers are accelerating relative to each other in different reference frames, instead of moving at a constant speed? The general relativity theory answers this question. The additional postulates of the theory of general relativity, which considers relative motion with uniform acceleration rather than the uniform speed, are as follows:

3. *If a frame K' uniformly accelerates instead of moving at a uniform speed relative to a Galilean (stationary) frame K, then we can*

*consider K' to be at rest by introducing the presence of a uniform
gravitational field relative to it.*

4. The laws of nature are the same in every frame of reference.

The third postulate is a simple consequence of the principle of
equivalence. It implies that an accelerating observer can consider
himself equivalent to a stationary observer, who feels the force of
gravity. This principle simply states that the gravitational mass
(measured by gravitational attraction) equals inertial mass (measured by
resistance to acceleration).

Einstein grasped this equivalence idea when he realized that a freely-
falling body in a gravitational field is the same as one that's moved by an
inertial force at an acceleration that equals gravity. We know that all
objects accelerate at the same rate in a gravitational field. Therefore, the
effect is the same as if the uniformly accelerating frame is at rest and that
a uniform gravitational field is present.

The fourth postulate is simply a generalization of the first postulate, as
it states that the laws are same, not just in an inertial frame, but in every
frame of reference. By supposing that K' is at rest and a gravitational
field is present in the third postulate, we make K' a Galilean frame,
where we can study the laws of mechanics.

A consequence of general relativity is that a gravitational field also
influences time and space. Scientists have measured time dilation due the
presence of a gravitational field. In fact, scientists have measured   and
found that the value of the gravitational field at the top of a high
mountain is a little less than its value at the bottom. Thus, two atomic
clocks initially synchronized gave two different results after having spent
a while in these two different places.

Another strange consequence of the above postulates of relativity is
that is Euclidean geometry cannot describe the world we live in.
However, another kind of geometry, developed by Gauss and later by
Riemann, can describe it. In this geometry, the circles are not round;
parallel lines can cross or diverge, and the angles of a triangle may not
add up to 180°.

- Einstein & Gravity

It is interesting to note that the gravitational force, which keeps us all
glued to the surface of Earth, plays an important role in the evolution and

fate of the Universe. Gravity ranks as the weakest of the four fundamental forces in nature (the other three being the electromagnetic, strong and weak nuclear forces. We shall discuss these forces soon). However, fields other than gravity, though much stronger, are not as important as gravity when we talk about the Universe as a whole. The weak and strong fields act only over very short distances inside the atoms; most large objects are electrically neutral, and thus unaffected by the electromagnetic field. However, the gravitational field affects everything at any distance.

Einstein redefined gravity because he was uncomfortable with the Newtonian concept of gravity, which could affect objects at a distance instantly. Einstein developed the general theory of gravitation along with the general relativity theory. Einstein modified Newton's view of gravity and argued that the curvature of space-time is a measure of gravity, as gravity is responsible for this curvature2. According to Einstein, the idea of curvature – somewhat similar to the curvature of a large sphere-like ball - also applies to the space-time geometry and the whole Universe. Gravity curves the space-time geometry.

According to Einstein's theory, the force of gravity in terms of the space-time curvature is somewhat similar to the force of attraction between the leaves floating in a still pond. When a leaf lands on the surface of a still pond, just like an object in space, it creates a tiny disturbance. This disturbance spreads with a certain speed across the surface as a ripple, deforming the surface of the pond, as shown in Fig. 1.7. Another leaf in the pond, like an object in space, also creates a similar slight depression in the smooth surface of the pond. Then, the two leaves, like two objects in space, attract each other, just as two persons would on a bed that sags.

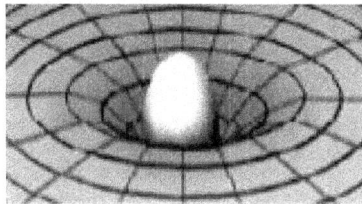

Fig. 1.7 – Illustration of space-time curvature due to an object

Einstein's theory of general relativity thus predicts the curvature and ripples in space due to gravity. In fact, it predicts that the detonating stars and the colliding black holes would rattle the space continuum and cause vibrations in the fabric of space. These ripples should expand outward at the speed of light, stretching and squeezing space, and cause the distance between objects to expand and contract. In other words, such events should generate gravity waves.

Unfortunately, by the time such waves reach Earth, the vibrations become so weak that they alter distances by less than one part in a trillion billion. As stated by Gibbs in an article in the April 2002 issue of Scientific American, detecting such a minuscule quiver is like detecting Saturn moving closer to the Sun by the width of a hydrogen atom ($\sim 10^{-10}$ m). Thus, we have not been able to directly detect the gravity waves.

Nevertheless, scientists have spent eight years, and $365 million, building an observatory to detect gravitational waves. A German/UK team has set up the giant GEO 600 gravitational wave detector and put it in a continuous observational mode. The Hanover lab is trying to detect the ripples created in the fabric of space-time. These ripples sweep out from merging black holes or exploding stars.

If successful, it would confirm fundamental physical theories and open a new window on the Universe. It would enable scientists to probe into the moment of creation itself. Compelling independent corroboration could also come from a spacecraft that can see the burst of gamma-ray radiation, which accompanies the cataclysmic events that produce gravitational waves.

Einstein's theory essentially states that the curvature in space-time directly relates to the energy and momentum of everything in space-time, except space-time itself. We call this 'curvature gravity'. Everything else in the Universe - namely, all the fundamental particles comprising matter and radiation, etc. - determines the curvature of space-time. In simple terms, an object brought closer to another object affects the curvature of space-time between the two objects, as if creating a dent, which results in the gravitational force attracting these two objects.

According to this theory, gravity is described by the curvature of space-time, tensor $G_{\mu\nu}$, which is proportional to the energy and momentum tensor, $T_{\mu\nu}$[3]. The basis of this elegant mathematical description in terms of differential geometry is the equivalence principle. Experimental confirmations of this theory are the bending of light by Sun, perihelion precession of Mercury, and the spin down of binary pulsars from gravitational radiation.

It is interesting to note that even a scientist of Einstein's caliber made many errors while developing the general theory of relativity and the gravitational theory. Einstein developed the final version of general relativity after a slow progress, with many errors along the way. In December 1915, he said about himself,

"That fellow Einstein suits his convenience. Every year he retracts what he wrote the year before." Most of Einstein's colleagues were at a loss to understand the quick succession of papers, each correcting, modifying and extending what he had done earlier.

Einstein also developed the correct equation of motion for describing a Universe. It relates the curvature of space-time in a given Universe to the distribution of energy and momentum in that Universe. The energy-momentum tensor, $T_{\mu\nu}$, includes all of the energy, from all the non-gravitational sources such as matter, electromagnetism, or even quantum vacuum energy, which we'll discuss later. This equation is the starting point for most of the modern developments in cosmology.

Finally, we must note that Einstein's theories of relativity deal with the macroscopic behavior of the Universe; quantum mechanics deal with microscopic behavior. Unfortunately, the relativity theories and quantum mechanics are incompatible.

1.6   Space, Time, Reality

Our understanding of reality, of time and space at the fundamental level, despite the theory of relativity, is far from perfect. We know that space and time are two of the most important elements of the Universe. However, we still do not have a complete understanding of their true nature. Space and time are important since change - the main characteristic of every phenomenon in this Universe - occurs in space and time. Change occurs according to the principle of causality, which specifies change because of some cause in terms of space and time parameters. In addition, we express the energy transformation accompanying change as a function of space and time parameters.

In fact, most of the phenomena in this Universe involve space and/or time parameters. Space, time, and causation, which enable our mind to perceive reality, cannot exist by themselves. Space has to relate to some object; the notion of time comes about for connecting two events by the idea of succession, and causation is expressed in time and space. It is important to revisit space, time, causality, and reality. They do not have independent existence, and yet they exist in the sense that we observe the Universe, and the changes in it, through a combination of these elements.

Our view of reality is also imperfect, because of the limited range of sensors that we can use for observation. Some even say that the physical world around us is an illusion of our senses. Our senses perceive only a part of reality. For example, our eye can only perceive a small portion of reality, as it fails to see over 90% of the remaining matter. Furthermore, our eye-lens decides the shape of the objects that we see.

Our brain's capacity to process the perceived information is limited, and we use a mere 4% of our brain's capacity, according to medical science. An ant, a bird, a cat, an elephant, and all other living entities have a different perception of reality. What is the absolute reality of the Universe, irrespective of the observer and the limitations of the sensors used to measure the reality in space and time?

Regarding time and space, the laws of science consider both space and time as symmetric parameters. We think of three-dimensional space as symmetric as it stretches in all directions, and think of the fourth dimension - time - as asymmetric, since we cannot go back in time. Why is the time dimension different from the space dimension? Could there be higher dimensions for space or even time? Is space a continuum or discrete? What is quantum space? Such problems need to be resolved about time and space

Scientists have difficulty in coming up with a clear definition of time. Time is the separation between action and reaction. We may define it as the distance between cause and effect. It enables an observer to order events in a sequence. However, many scientists envisage time as being there all at once, a 'timescape', stretched out like a landscape. We often refer to this concept as the 'block time'. In block time, we treat time simply like a coordinate to label events, and associate different values of coordinates in the reference system with past, present, and future, just as we do on a spatial axis. Some believe that in reality, the past, present, and future are always with us. The limitation of our senses simply prevents us from perceiving them. The past appears to be gone, and the future is not yet here.

Suppose we divide time into three parts, namely, the 'past', which is gone and no longer exists; the 'present', which is now and is reality; and the 'future', which is yet to come and does not exist now. The present time, or now, thus keeps flowing forward. As now marches or flows forward, it transforms future events into the new present reality, and what was the present reality slips into the past. We cannot define time in the past because the past does not exist anymore, and we cannot define time in the future because it too does not yet exist.

What do we mean by the passage of time, and how do we measure it? We normally observe an event, such as the motion of needle on a dial or oscillations in an atomic clock; as its position changes by a certain amount, we associate that with a passage of one second of time. Actually, we are measuring the change of position in this case and not the passage of time.

The question as to how fast time moves or passes is absurd. After all, how fast does time pass, and what does it pass (or move) with respect to? In fact, what we are observing is the change in the state of an event (e.g. position of an object) and not the passage of time. We can say that a car is moving at eighty km/hr, but when the object we measure is time itself, what do we use as a measure? A 'second per second' does not make any sense.

Until the beginning of the twentieth century, everyone thought that both time and space were absolute and one could represent an event in time or a distance in space by a fixed number. As discussed in the preceding section, Einstein then came up with strange conclusions about time and space. He said that the measure of time and distance is different for different observers.

According to his special theory of relativity, when two observers move relative to each other at constant speed, they get a different measure of time and distance when observing the same events. Einstein thus demolished the notion of absolute space or time, since each observer has his own measure of space and time.

How do we define 'now'? Einstein's theory of special relativity says that we cannot attach any absolute and universal significance even to 'now' or the present. Einstein showed that two spatially separated events judged to occur simultaneously by one observer could occur at different moments for another observer. Thus, Einstein demolished the idea that time is universal with a common present, or 'now', for everybody. It becomes hard to define time when the notion of simultaneity becomes relative. As mentioned earlier, Einstein himself said that the past, present and future are only illusions, however persistent. In short, it is hard to define 'now' and the passage of time in absolute terms.

Einstein also said that time is like a flowing river. Just as in a river, the more mass or energy an object possesses, the more current varies around it. The more mass, concentrated at a point in a flowing river of time, bends the flow of time more. However, when we talk of the passage of time, we think of it as flowing like a river in one direction - what we call an arrow of time - from past to present and to future.

To add to the present confusion, Kaku writes in the August 2005 issue of the Wired Magazine that it is theoretically possible to divert the river of time into a whirlpool – a closed time-like curve, or a fork. Such a fork would lead to a parallel Universe, and it could make possible a basic time machine for time travel

The passage of time and the so-called arrow of time happen because events in this Universe form a unidirectional sequence. We think of the arrow of time pointing towards the events to follow in the future. Just because this arrow points towards future events, it does not mean that the arrow is moving towards the future.

The arrow of time just denotes the asymmetry of time, unlike space, which is symmetric. If our perception of the flow of time is a mental quirk, rather than a property of the physical Universe, then what causes it? Some think that it is the language structure; others attribute it to the workings of the brain.

As regards the perception of the flow of time by the brain, it could be due to two possible reasons. We discussed the second law of thermodynamics, which implies that the entropy of a closed system always increases in any change; an enclosed system must move from a more ordered to a less ordered state. Most of the processes in nature are irreversible, and the second law of thermodynamics creates asymmetry, as entropy increases with time. This unidirectional increase in entropy in our brain is somehow linked to the perception of the flow of time. That is, because of the second law, we usually recount history forward.

The second possible reason for our perception of the flow of time comes from quantum mechanics, as suggested by Penrose. Heisenberg's Uncertainty Principle implies that nature is inherently non-deterministic and we cannot predict either the future or the past. Quantum mechanics, discussed in detail in the next section, implies that a quantum state has potentially many possible alternative outcomes with different probabilities, which one can compute. However, when we observe, we observe only one definite outcome.

The act of observation projects one outcome, or 'reality', out of several possible outcomes. We thus observe what we call 'reality'. In other words, a transition takes place in our brain from the many possible outcomes to a single outcome, or from an open future to a fixed past, which is what we call the 'flow of time'.

Nevertheless, we normally think of time flowing like a river in one direction – from the past to the future, through the present. Time does point in a definite direction, and one can trace the origin of time's arrow back to the Big Bang. We've discussed another point of view that time does not flow. The act of observation simply gives us a perception of time flow. We also presented plausible physical reasons why we get this perception. Some scientists even talk about two dimensions for time. Hawking provides an interesting discussion on the notion of time in his books, and he even advances the notion of imaginary time.

Hawking also points out that, unless we can run time backward, we cannot understand the biggest mystery of the Universe. In a recent talk in 2007, Hawking calls it a 'top-down' approach to what he has long considered the biggest cosmic question: What was the initial state of the Universe? Did God create the Universe the way it was, and that is it? Alternatively, is there a scientific reason for why the cosmos is just so? Why, for instance, did it lead to the conditions for intelligent beings like us?

"We say that later events are caused by earlier events, but not that earlier events happen in order to lead to the later." "We do not know what the initial state of the Universe was, and we currently can't try out different initial states and see what kinds of universes they would produce."

This 'bottom-up' approach, as Hawking called it, works well in situations in which we can choose the initial state and observe the outcome. However, the bottom-up approach does not work in cosmology. General relativity alone cannot solve the problem. Therefore, quantum mechanics has to come into play to figure out the likeliest backward history for our Universe.

Getting away from these weighty questions, let us pose the following question: If we isolate a person completely from the environment, such that his senses do not receive any outside data, and the mind does not process or observe any such data, does the person lose all sense of time and space? It seems that his biological clock keeps ticking as he keeps processing the data received from the internal parts of the body. However, he feels disoriented in time and space. It seems that one needs a reference point in time (space) and some means to measure the passage of time (distance) for proper orientation.

Another interesting question would be as follows: what is the precise role of the observer in the definition of space and time? If there were no observers left in the Universe to observe or measure distance and time, what happens to space and time? In this Universe, events continue to take place in space and time, whether there is someone to observe them or not. The nature of time and space changes as soon as an observer attempts to measure them. The relative position of an observer and the act of observing the effect of measuring time (or space) can affect the effect that we are trying to measure.

Nevertheless, we do need an observer to get a measure of space and time. Einstein's theories talk about observers in different reference frames moving relative to each other, carrying measuring rods and clocks, to measure the shortening of measured lengths and observe the slowing of clocks. The distances perceived by each moving observer relative to each other shorten and time slows down, since the speed of light (distance/time) in free space is unaffected by motion.

While discussing classical and quantum mechanics, time enters the picture in an interesting manner. According to Kauffman and Smolin in their paper, "A Possible Solution for the Problem of Time in Quantum Cosmology", time (t), in ordinary classical mechanics, refers to a clock carried by an inertial observer. It is not part of the dynamical system, which one models. One uses this clock to label the trajectory of the system in the configuration space. However, for the external clock, one could already say that time has disappeared, as each trajectory exists all at once as a curve (g) in the configuration space. Once we choose the trajectory, the whole history of the system is determined.

In this sense, nothing in the description corresponds to the common thinking of the flow or progression of time. The whole trajectory exists when one specifies any point and velocity. Similarly, we could say that the whole set of trajectories exists as well, as a timeless set of possibilities. We represent time in the description, but it is not associated with the system itself. It may be said that time is not something physical, as represented in classical mechanics. In quantum mechanics, the situation is rather similar. There is time (t) in the quantum state and in the Schrödinger's equation, but it is time measured by an external clock, and is not part of the modeled system.

From the preceding discussions, it is obvious that we need a better understanding of the notion of time. Do we accept Einstein's assertion that that the past, present and future are only illusions, however persistent, and envisage time as all there at once, like a landscape? Do we treat time simply as a coordinate to label events, just as latitude and longitude are used to label places? In addition, do we treat our perception of the flow of time as simply a state arising from the manner in which our consciousness, or our brain, resolves ambiguous quantum states? We shall revisit time and space when we discuss the frontiers of science in the third chapter.

1.7   Causality

Causality, along with space and time, essentially defines the Universe as we know it. The causality principle simply states that there can be no effect without a cause, and the effect cannot precede a cause in time. The cause, or an action, starts something off, and then one observes the effect ensuing from the action as a function of time. Causality thus relates to the process of change. We know that energy is the prime cause of change. Thus, the interaction of force fields and the exchange, or transformation of energy, causes the change. Since energy cannot travel faster than the speed of light, the effect cannot manifest faster than this limit.

Science essentially seeks the cause of every phenomenon in this Universe, since it believes that every effect has a cause. Since every effect has to have a cause, and science cannot find the ultimate cause for the origin of Universe, then religion steps in with the belief that the Creator created the Universe. Of course, religion does not answer the question as to who created the Creator. This is a dilemma. In fact, every effect becomes a cause that causes another effect, and the chain continues. Thus, one can never find the ultimate cause, because the question would always arise, 'What caused the ultimate cause?'

The cause and the effect are essentially the same, since an effect becomes a cause for a subsequent effect. Since the process of change in the Universe is always ongoing and eternal, one observes the cause-effect phenomenon continuously. One way to resolve the preceding dilemma is to assume that the process of cause and effect is a closed chain, and every effect becomes the cause in a loop. Thus, there is no beginning as the end is the start of a beginning. We must also note that the presence of an object in the Universe affects the other objects and the space itself. Everything in the Universe is interrelated, interdependent, and causally connected.

We think that every cause has a definite and not random effect, governed by the deterministic laws of nature. We also think that the relationship between cause and effect has to do with the object only, and with the physical laws it obeys. We normally think it has nothing to do with the observer, or the observation process, once an action is set in motion. Quantum physics changes all that. As stated, mere observation is enough to alter the history of an event.

Quantum mechanics also introduces the element of uncertainty in the process of change. One might also ask the question, 'Can we always observe and control every state of a system?' In fact, modern control theory defines the concepts of controllability and the observability of a state of a control system, which might provide answers to these questions.

Is the causality principle ever violated, and if so, why and when? Some scientists are even beginning to doubt the principle of causality. Some recent experimentalists claimed that, by suitably pre-adjusting properties of a system, they could obtain a response at the other end at a speed faster than that of light.

In 1935, Einstein Podolsky and Rosen proposed the following dilemma through mathematical reasoning: If quantum theory were correct, then a change in the spin of one particle in a two-particle system would affect its twin simultaneously, even if the two had been widely separated in the meantime. The Einstein-Podolsky-Rosen (EPR) paradox caused a stir about causality. It also posed a dilemma for quantum theory. The theory of special relativity forbids simultaneous effect with cause, since a signal cannot travel faster than the speed of light.

In 1964, Bell gave a theorem essentially stating that, if the statistical predictions of quantum theory were true, then an objective Universe would be incompatible with the law of local causality. It implied that EPR's seemingly impossible proposition did hold true and instantaneous changes in widely separated systems did occur. The implications of Bell's Theorem pose a great problem for scientists. How can two particles, once in contact, separated even to the ends of the Universe, change instantaneously when a change in one of them occurs, unless they are still connected or entangled in some sense?

To explain this strange behavior, Sarfatti of the Physics/ Consciousness Research Group proposed that no signal, requiring energy, is transmitted between the distant objects, but only the 'information' is transmitted. Thus, one does not violate Einstein's special theory of relativity. However, it is not clear as to what exactly this 'information' is and how it travels instantly without requiring any energy.

D'Espagnat thinks that all objects constitute an indivisible whole. Stapp suggests that it would be a mistake to suppose that these effects operate only with relevance to the invisible world of the atom. He states that these findings imply that the oneness, implicit in Bell's Theorem, covers human beings and atoms alike. It is obvious from these discussions that a complete understanding of causality is very important, since it is responsible for change in this dynamic Universe.

- Is the Universe a Hologram?

Some quantum physicists now say that each part of the Universe contains all the information present in the entire cosmos itself, similar to a hologram or a seed that contains all the information to replicate it. A hologram is a specially constructed image from the interference of a direct laser light beam, reflected from a 3-D object. This hologram, when illuminated by a laser beam shows the same 3-D object suspended in space. The most incredible feature of holograms is that any piece of it can provide an image of the entire hologram, although with less detail. The basic information of the whole is thus contained in each part.

According to Bohm, the Universe is similar to a hologram. Each part contains within itself the form and structure of the entire Universe. Each part of the Universe contains enough information to reconstitute the whole. According to Pribram, brain photographic film has the information of the Universe encoded. The brain encoding information is a hologram that is a part of an even larger hologram - the Universe itself.

Some scientists apply indivisibility to space and time, since relativity shows them to be inextricably linked, interconnected, and inseparable. Bell's Theorem involves the non-local features of the Universe: objects once in contact, though separated spatially, even if placed at distant ends of the Universe, are somehow in inseparable contact. These non-local and non-causal descriptions are for objects separated in space. Bohm states that the implications of quantum theory apply to moments in time also. We normally think that quantum physics applies only at a microscopic level in nature - electrons, protons etc. - and that relativity has only to do with massive objects of cosmic proportions - stars, galaxies, nebula etc. Thus, according to Bohm, one has to think of the entire Universe as a single undivided whole.

Some suggest that the interrelation of human consciousness and the observed world is obvious in Bell's Theorem. We cannot separate our own existence from that of the world outside. We are intimately associated, not only with the Earth we inhabit, but with the farthest reaches of the cosmos. Thus, one cannot regard human consciousness and the physical world as distinct and separate entities. Our thoughts shape, to some extent, the reality of the external world.

Deepak Chopra, at Stanford University, recently attempted to demonstrate that conscious mental activity exerts measurable effects on the physical world - a world that includes human bodies, organs, tissues, and cells. The mind influences our health and sickness. According to these claims, the dividing line between life and non-life is illusory and arbitrary. According to Hindu scriptures, no division is admissible, even in consciousness at any time, and the individuality of the conscience is false. After you take away a part, the remainder is still complete.

Regarding the link between quantum physics and consciousness, we must note that modern physics does not recognize consciousness as a separate element, like matter or energy. This is mainly because physical science deals with the study of inanimate matter, energy, and the related laws of nature. Recent studies in life and medical sciences, related to the human brain, do attempt to define consciousness, which gives each individual a personality and makes us aware of the external world.

Many people usually associate conscience with the soul, bringing into the discussion the issue of religion when talking about conscience. However, this need not be so. For example, one can define consciousness or awareness as the projection of an image by the brain through different neural connections, based on the data provided to the brain by our physical senses. Lest we venture further into the controversial domains of philosophy and religion, let us conclude this first phase of our journey in which we have unfolded the mysteries of science.

1.8 Modern Physics – Amazing Discoveries

These days, we simply cannot escape talk about atoms, fission (atom bomb), fusion (hydrogen bomb), X-rays, nuclear radiation, radiation therapy, magnetic resonance imaging (MRI), computer-aided-tomography (CAT) scans, chips, integrated circuits, etc. These terms are part of the jargon of modern physics, which are responsible for the marvels of modern technology.

Let us step into the field of modern physics, and learn what we can about it. We'll learn that energy comes in fixed quanta, like different coins. We'll learn about an atom, which is the basic element of matter. We'll look inside an atom, and learn how science discovered the fundamental structure of an atom. We'll learn about quantum physics. We will then visit the important historical developments leading to the atomic and the Standard Model for particle physics.

It all started with a problem with the classical notion that energy, emanating from a blackbody, is continuously distributed radiation. A blackbody absorbs all radiation and does not reflect any radiation incident upon it. Because it reflects no light, it would appear black to an observer. It is also an ideal radiator. The simple calculations based on classical physics showed that the energy radiated from the blackbody should be infinite, which it is not. Planck resolved this problem by conjecturing that energy is packed in discrete packets, called quanta.

The idea of quanta of energy inspired Einstein to develop the photoelectric equation. In photoelectric phenomenon, light impinging on an atom knock off electrons from an atom, and these electrons produce an electric current. It inspired Bohr to develop the atomic model for the hydrogen atom, with an electron (negatively charged atomic particle) revolving in a certain orbit around a proton (positively charged particle in the nucleus of an atom). It inspired de Broglie to propose the wave nature of the electron and matter, relating mass and momentum to the wavelength and the frequency of the waves. It finally led to the development of Schrödinger's wave equation, quantum mechanics, and the Standard Model for the atom.

A. Quanta of Energy

Scientists often observe a relatively insignificant phenomenon, and discover important laws of nature. First, Newton discovered gravity, when he observed an apple fall to the Earth. Einstein discovered relativity theories, based on his discomfort with the asymmetry between electric and magnetic phenomenon, and with the notion of gravitational field propagating instantly. Another example is Planck's discovery of quanta, on noticing a problem with the classical notion of continuous energy radiating from a blackbod, as shown in Fig. 1.8.

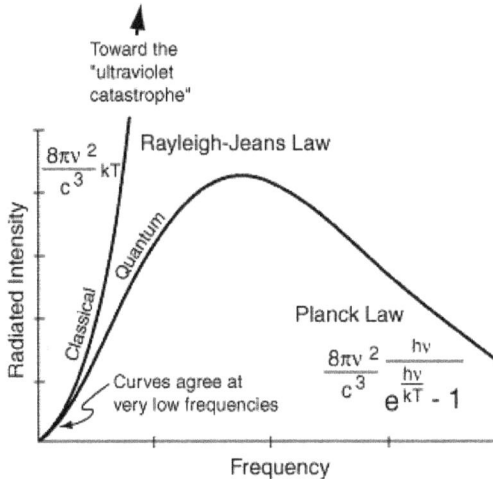

Toward the "ultraviolet catastrophe"

Rayleigh-Jeans Law

$$\frac{8\pi v^2}{c^3} kT$$

Radiated Intensity

Quantum

Classical

Curves agree at very low frequencies

Planck Law

$$\frac{8\pi v^2}{c^3} \frac{hv}{e^{\frac{hv}{kT}} - 1}$$

Frequency

Fig. 1.8 - Blackbody radiation intensity equations

Regarding blackbody radiation, Kirchhoff proved, in 1859, a theorem which states that the energy emitted (E) from a blackbody depends only on the temperature (T) and the frequency (v) of the emitted energy, i.e., $E = J(T, v)$. Kirchhoff challenged scientists to find the exact form of the function J.

In 1879, Stefan proposed, on experimental grounds, that the total energy emitted by a hot body was proportional to the fourth power of the temperature. Boltzmann reached the same conclusion in 1884 from theoretical considerations, using thermodynamics and Maxwell's electromagnetic theory. The result, now known as the Stefan-Boltzmann law, did not fully answer Kirchhoff's challenge, since it only established the temperature but not the frequency relationship.

In 1896, Wien proposed a solution to Kirchhoff's challenge. However, Rubens and Kurlbaum showed that it breaks down in the infrared region. For blackbody radiation, the classical law formulated by Rayleigh and Jean gave an expression for the power spectral distribution function

In 1900, Planck guessed the correct formula for Kirchhoff's J-function. Although Planck's guess fitted with experimental evidence at all wavelengths, he was not satisfied and tried to develop a theoretical derivation of the formula. To match the observed data, Planck said that the energy of electromagnetic radiation comes in packages, or quanta, like the fixed denominations in currency.

One can obtain Planck's expression by multiplying the density of states in terms of frequency or wavelength with the photon energy and with the Bose-Einstein distribution function with normalization constant as unity. Planck asserted that the energy (E) of one package is proportional to the frequency of radiation (f). $E = h f$. We call this proportionality constant (h) the Planck's constant with a value 6.6262×10^{-34} J-sec (4.136×10^{-15} eV.s). Finally, Planck derived the expression, shown in Fig. 1.8 that fits in excellently.

The important implication of Planck's assertion is that the energy emission is not continuous, and a group of atoms cannot emit an arbitrarily small amount of energy at a very high frequency. In other word, only atoms (oscillators) with a large amount of energy can emit at high frequency. Since the probability of finding groups of atoms with such unusually large energies is low, the spectral curve falls at high frequencies or short wavelengths ($\lambda = c/f$). The quantum view expressed in the Planck hypothesis is that, either you add the energy of a whole photon, or you do not add any at all.

One also observes the discrete levels of energy in the discrete spectrum for various elements. We observe that the spectrum of light emitted by energetic atoms is composed of individual lines of different colors. We now know that these lines represent the discrete energy levels of the electrons in those excited atoms. If electrons were not restricted to discrete energy levels, the spectrum from an excited atom would be a continuous spread of colors from red to violet, with no individual lines.

When an electron in a high-energy state comes down to a lower one, the atom emits a photon – a carrier of an electromagnetic field. The emitted energy corresponds to the exact energy difference between these levels (conservation of energy). The larger the energy difference, the more energetic the photon will be, and the closer the spectral line would be to the violet end of the spectrum.

B. Einstein & Photoelectric Effect

Solar cells, used in many applications, convert light from the Sun into electrical energy. Einstein enunciated the basic principle for this phenomenon in 1905. He applied Planck's idea to explain the photoelectric effect, in which the metal surface, exposed to light radiation, emits electrons. He assumed that light energy comes in small discrete bundles. It is not distributed continuously. He called these discrete bundles of energy 'photons', each having an energy, $E = h \, f$. He speculated that the metal atoms absorb light quanta up to a trigger frequency threshold. Above it, the quanta energy frees the peripheral electrons - no longer bound to an atom - which are then emitted.

Einstein's basic idea is quite simple. Suppose the frequency of the incident light is high enough so that energy (E) exceeds or equals the characteristic energy, called the work function of the particular metal, ($\phi = h \, ft$), corresponding to the threshold frequency (ft). Then the metal atoms absorb the light quanta, and the energy quanta free a peripheral electron, no longer binding it to the atom.

Einstein thus developed the Photoelectric Equation.[4] It states that the difference between the incident energy (E) and the work function (ϕ) equals the maximum kinetic energy that a free electron with mass (m) can acquire [$E - \phi = (mv^2/2)$ max]. Some electrons could have less because of the loss of energy from traveling through metal. Einstein was awarded a Nobel Prize for this work related to quantum theory, and not for his relativity theories. It is ironic, because Einstein never quite believed in some of the random aspects of quantum theory.

C. Atomic Model

What constitutes matter, and how far could we go if we kept dividing all matter into smaller parts? Aristotle believed that only four basic elements make up all the matter in this Universe. According to Greek and ancient Hindu philosophy, five basic elements (Bh-a-ga-va-n) constitute the five essential elements of the Universe, namely, Earth (Bhu), fire (Agni), space (gagan) air (vayu), and water (neer) Universe. All these elements are also inseparable parts of the human body.

In 1803, Dalton said that if we keep dividing matter into smaller and smaller pieces, we would we end up with atoms. He used the word 'atom' (Greek for 'indivisible') and showed that the grouping of atoms form molecules. Could we split the atom further? Thomson answered this question as "yes", when he demonstrated the existence of electrons inside an atom.

Rutherford and Bohr proposed an atomic model. According to this model, the nucleus of an atom contains positively charged protons, and negatively charged electrons revolve around the nucleus. Can you believe that they discovered electrons only a little over 100 years ago, in 1897? Thomson did not expect it to be a major event, as illustrated by his remarks. He said,

"I was told long afterwards by a distinguished physicist who had been present at my lecture that he thought I had been pulling their leg."

The number of electrons revolving in various orbits around the nucleus equals the number of protons in the nucleus, which keeps the atom neutral.

It was recently discovered that an electron in the hydrogen atom completes its orbit around its nucleus in 150 attoseconds (one attosecond = 10^{-18} seconds). Shown in Fig. 1.9 is an illustration of the hydrogen atom, whose size is extremely small (around 10^{-12} m).

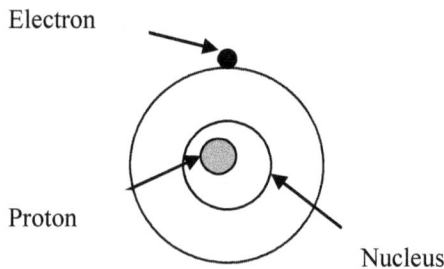

Fig. 1.9 – Illustration of Hydrogen Atom

How far we have come in these hundred years! As reported in early 2008 by Mauritsson of Lund University in Sweden, scientists have filmed an electron in motion for the first time. As stated, it takes about 150 attoseconds for an electron to circle the nucleus of an atom. An attosecond is 10^{-18} seconds long, which relates to a second as a second relates to the age of the Universe. The technique used for recording the motion of an electron uses extremely short flashes of light, which are necessary to capture an electron in motion. A technology developed within the last few years can generate short pulses of intense laser light, called attosecond pulses, to get the job done. This new technique, according to Mauritsson, will allow researchers to study the tiny particle's movements directly.

Returning to our journey, in 1932, Chadwick discovered that the nucleus could also contain another type of particle, the electrically neutral 'neutron'. An atom is mostly empty space except for protons and neutrons. When one shoots subatomic particles at a very thin gold foil, most of the particles go straight through the gold. Almost all the mass of an atom is concentrated in the tiny nucleus. It is not possible for anyone or any machine that uses light to actually see a proton. The wavelength of light is too large to detect it (2.10×10^{-16}m).

It is interesting to note that all protons, neutrons, and electrons have exactly the same characteristics in the atoms of all elements. Protons and neutrons have almost exactly the same mass, and they are all inside a nucleus in the center of an atom. Electrons orbit outside the nucleus in electron shells, which are different in shape at different distances from the nucleus. The mass of an electron is 9.05×10^{-28} gm, which is about 1/1835 the mass of a proton (1.66×10^{-24} gm. or 1 AMU). Electrons have a unit negative charge and protons have a unit positive charge. These charges are genuine charges. Neutrons do not have any charge.

- Questions about the Atomic Model

During the various stages of the development of the atomic model, several questions and problems arose, such as the following:

1) What keeps the negatively charged electrons apart from the positively charged protons? If electrons were stationary, then protons would attract the electrons due to Coulomb force. This problem forced physicists to assume that electrons were not stationary, but revolved around the nucleus, generating enough centripetal force to counteract the Coulomb attractive force.

2) The circling electron should radiate its energy away, like microscopic radio antennas, and spiral into the nucleus. A negatively charged electron, moving around a positive nucleus, is an oscillating dipole. When one views an atom in the plane of

orbit, the negative charge appears to oscillate up and down relative to the positive charge. Such an oscillation would transmit electromagnetic waves, the atom would lose energy, and the electron would spiral into the nucleus. This problem forced Bohr to postulate in 1913 that electrons could move only in non-radiating orbits, called stable stationary states (like standing waves in a fixed string) with a certain amount of fixed energy. An atom, however, does radiate energy (ΔE) when an electron makes a transition from a high-energy stationary state (E_i) to a low-energy stationary state (E_f). The frequency of radiation (f) is not the frequency of electron's motion in either stable orbit. It is related to the energies in initial orbit and final stationary orbits as follows: $(E_i - E_f) = \Delta E = hf$.

3) The spectral lines of the hydrogen atom when excited by an electric discharge and observed through the emission of light by hydrogen atom were discrete. The spectrum indicated that the frequency (f) (or wavelength (λ)) of the radiated light was related to the square of integers. The problem for the model was to find a physical explanation for the spectral lines. Bohr had postulated the stable stationary orbits for electrons. He also suspected that the radii of these orbits should relate to the square of the integers. He succeeded through trial and error. He postulated that the angular momentum (m.v.r) of an electron of mass (m) moving with velocity (v) in a stationary orbit of radius (r) equals an integer (n) times the Planck's constant divided by 2π.

Many questions remained unanswered about this atomic model. For example, why do electrons occupy so-called non-radiating stable stationary orbits with fixed amount of energy? To answer this question, scientists went on to develop quantum mechanics, essentially a set of mathematical rules, as we shall see shortly. Other questions concern the true nature of fundamental particles. Are they real particles or some form of wave fields? At the macroscopic scale, we are used to two broad types of phenomena: waves and particles. The particles are localized phenomena that transport both mass and energy as they move. The waves are non-localized phenomena, as they spread-out in space and carry only energy but no mass as they propagate. To give a simple example, physical objects that one can touch are particle-like phenomena. However, ripples on a lake, which do not transport net water mass, are waves.

Debate on the nature of fundamental particles continues. Fundamental particles cannot extend in space. This is because different parts of an extended charged particle would exert a repulsive force against each other. This force would approach sufficient strength at these very small distances to tear the particle apart. The elementary particles with certain mass have to be Dirac point particles (named because of the Dirac delta function) with zero volume and infinite density. A geometrical point cannot spin. Dirac point-particles in an atom are also supposed to be moving at quite high speeds in certain orbits.

In quantum mechanics, scientists circumvent such questions through particle-wave duality. In string theory, they talk about the modes of vibrations. In quantum mechanics, this division between particles and waves is not that clear. The so-called particles, such as electrons, can behave like waves in certain situations, while waves, such as electromagnetic radiation or light, can behave like particles, photons. Electrons, therefore, can create wavelike diffraction patterns upon passing through narrow slits. On the other hand, one can explain the photoelectric effect only if light behaves like photon particles. Such ideas led de Broglie to conclude that all matter has both wave behavior and particle behavior, known as the principle of wave-particle duality. Let us continue our journey towards quantum mechanics by visiting de Broglie's ideas about matter.

D. Wave Nature of Matter

In 1924, Louis de Broglie suggested in his doctoral dissertation that electrons might have wave properties. He was the first person to propose the wave nature of all matter[5]. However, the wavelength of normal-sized objects moving at normal speeds turns out to be incredibly small. It is so small that no one can notice the quantum mechanical effects at the macroscopic level. The propagation of waves of a very small wavelength is indistinguishable from the propagation of particles.

The wavelengths (λ), calculated from de Broglie equations ($\lambda = h/p$, p – momentum) for macroscopic objects, are extremely small. Hence, we do not observe the wave properties of matter at the macroscopic level, and Newtonian mechanics is perfectly acceptable for everyday applications.

On the other hand, small objects like electrons have their wavelengths comparable to the microscopic atomic structures they encounter in solids. Thus, a quantum mechanical description at the microscopic level, which includes their wavelike aspects, is essential to their understanding.

Thomson's observation of electron diffraction pattern, in the transmission of electrons through thin metal foils, did confirm the wave nature of electrons, which was normally considered particles of matter with certain mass, volume, and charge. De Broglie pointed out that Bohr's equation for the angular momentum of the electron in a hydrogen atom is equivalent to a standing wave condition. If energy is associated with the frequency of a standing wave, then the standing waves in different orbits of an atom imply different levels of energy.

E. Quantum Mechanics

What is quantum mechanics and why is it hard to believe? When we think about atoms, electrons, protons and neutrons, we are looking at the Universe at a microscopic scale. Quantum mechanics is mainly concerned with phenomena at this scale. Von Neumann, one of the great contributors to quantum mechanics, once said,

"Einstein's relativity theory was at least understood by Einstein and by several other people after reading the papers again and again. On the other hand, I think I can safely say that nobody understands quantum mechanics."

Others have called quantum mechanics Alice-in-Wonderland physics, with all of its seeming absurdities. Despite such comments, quantum field theory has proved to be an amazingly accurate description of nature. Quantum mechanics ignores gravity at the microscopic scale, and has the following basic postulates:

1) *Energy occurs in discrete bundles 'quanta' instead of continuously.*
2) *The subatomic particles have both particle and wavelike characteristics, obeying Schrödinger's wave equation6, which can determine the probability of the occurrence of an event.*
3) *Certain pairs of measurements have an intrinsic uncertainty associated with them.*

According to quantum mechanics, a particle can have two mutually exclusive properties at the same time, and an atom can simultaneously exist in different states, provided one leaves it undisturbed and unobserved. When we observe, the different possible states have to collapse, and the atom has to choose only one of the different possible outcomes. Quantum mechanics can predict the probabilities (or likelihood) associated with different choices. The outcome with higher probability is more likely to happen when observed. In other words, if we made the same measurement on a large number of similar systems, we could predict the probability of occurrence of a particular outcome out of several possible outcomes. By dealing with probabilities, quantum mechanics introduced an element of uncertainty into a deterministic world. It made the probabilistic approach a valid alternative to the deterministic one.

Several scientists have reservations about quantum mechanics. They consider it a set of mathematical formulae and rules that work. The equations of quantum mechanics work very well, although they do not seem to make physical sense. Nobody understands what really happens in terms of the physics at the microscopic scale. Bohr said,

"No one who is not shocked by quantum mechanics understands it".

Einstein was also dissatisfied with the reliance upon probabilities in quantum mechanics. He said, in 1926,

"Quantum mechanics is very impressive, but an inner voice tells me that it is not yet the real thing."

He was uncomfortable with the theory, and said that *God does not play dice with the world.* However, even more fundamentally, he believed that nature exists independently of the observer, and the motions of particles are determined precisely. He felt that it is up to physicists to uncover the laws of nature. He also believed that such laws, when discovered, would not require statistical theories.

As stated earlier, in 1935, Einstein Podolsky and Rosen came up with a strange conclusion based on quantum mechanics. Two particles separated at the opposite end of the galaxy happen to be 'entangled' in a quantum-mechanical sense, and one particle instantly 'feels' what happens to its twin. We shall discuss this non-locality phenomenon further when we visit the principle of causality. Even Schrödinger ridiculed the superposition concept in his theory by describing a half-living, half-dead cat. Quantum mechanics offers no clear reason for individual physical events, has no solid foundations, and provides no way to obtain intrinsic properties of an object.

Several scientists still believe that quantum mechanics is an incomplete theory. Einstein felt that quantum mechanics seem consistent only with statistical results and could not fully describe every motion. Several scientists claim that we can derive quantum mechanics from classical mechanics. In classical mechanics, particles have definite positions and velocities and obey Newton's or relativistic laws. The particles at a quantum level appear to have uncertainty about their positions or velocities, only because we do not or cannot observe the underlying order. Quantum mechanics, according to some scientists, has the same kind of randomness as a coin toss. A coin toss is random, but we could write a deterministic equation if we knew all the parameters including the initial condition.

On the other hand, this point of view has changed since Einstein's time. In fact, string theorists and other scientists, pursuing the grand theory to unify all the fields, think otherwise. They feel that Einstein's relativity theory, and not quantum mechanics, would prove to be a mere approximation. In other words, according to them, uncertainty and randomness is an inherent part of the Universe. In fact, Hawking states in his book, The Universe in a Nutshell, that according to available evidence, God is quite a gambler and the whole Universe is like a giant casino, where dice are rolled on every occasion.

Quantum mechanics introduces the concept of a wave function for particles, such as electrons. We know that, for a photon, the wave function has an intelligible physical meaning. It is just an electromagnetic wave. An electromagnetic wave, described by the Maxwell electromagnetic wave equation, is mere evidence of electromagnetic field oscillations. However, the situation is different for an electron. We understand what a particle behavior means for an electron, but electrons are waves of what field? These are neither electromagnetic waves nor an oscillation of any known physical field! So, what are they?

How can something be both a particle and a wave at the same time? For one thing, it might seem incorrect to think of light as a stream of particles moving up and down in a wavelike manner. Some scientists think of light as made up of photons, as discrete quanta of energy. The matter exists as discrete particles, such as electrons, etc. The wavelike behavior is that of the probability of where that particle will be and not of the particle itself. Light sometimes appears to act as a wave, because we notice the accumulation of light particles distributed over the probabilities of where each particle could be.

How can we explain the phenomenon of quantum tunneling, where a particle might 'tunnel' or leak through a barrier? We've stated, just now, that a wave determines the probability of where a particle will be. When that probability wave encounters an energy barrier, most of the wave is reflected back, but a small portion of it may 'leak' into the barrier. If the barrier has a small thickness, the wave that leaked through will continue on the other side of it. Although the particle does not have enough energy to get over the barrier, there is still a small probability that it can 'tunnel' through it! This is one of the most interesting phenomena to arise from quantum mechanics. Without it, the computer chip would not exist, and a PC would probably take up an entire room.

Suppose we throw a rubber ball against a wall. We know that the ball does not have enough energy to go through the wall, and we expect it to bounce back. According to quantum mechanics, however, there is a small probability that the ball could go right through the wall (without damaging the wall) and continue its flight on the other side! Fortunately, it turns out that, for something as large as a rubber ball, this probability is extremely small. Thus, we could throw the ball almost forever and never see it go through the wall. However, with something as tiny as an electron, tunneling is an everyday occurrence.

- Evolution of Quantum Mechanics

How did quantum mechanics evolve? As scientists developed a rudimentary model for an atom, and the ideas of discrete energy packets as quanta began to crystallize, we needed a theory to explain the behavior of atomic particles at the microscopic level. The idea of explaining the discrete energy states of the atom by standing waves, directly led to the development of quantum mechanics. Schrödinger related the frequency and wavelength of electron waves, defined by de Broglie, to the energy and momentum of electrons. The idea was similar to relating light waves to photon energy and momentum.

Schrödinger's central equation of quantum mechanics for the wave function, $\psi(r, t)$, cannot be derived in a satisfactory manner, and one must accept it as a postulate of quantum mechanics[7]. The approach suggested by Schrödinger was to postulate a function, $\psi(r, t)$, that would vary in both time and space in a wavelike manner. Within it, the wave function would carry information about a particle or system. The time-dependent Schrödinger equation allows us to predict the behavior of the wave function over time in a predictive manner, if we know its environment. The information concerning environment is in the form of the potential (V), experienced by the particle according to classical mechanics.

In 1932, von Neumann put quantum theory on a firm theoretical basis. Some of the earlier work had lacked mathematical rigor. Neumann put the whole theory into the setting of operator algebra. The imaginary number, i $(=\sqrt{-1})$ appears in Schrödinger's equation. Such an imaginary number, part of complex numbers, is extensively used in mathematics, science, and engineering. Hawking, in his book, The Universe in a Nutshell, recently used this imaginary number to introduce the concept of imaginary time to explain this Universe.

Scientists, including Schrödinger, wanted to know the physical significance of the electron wave function, $\psi(r, t)$. Born showed that the 'wave function' of Schrödinger's equation represents only the probability of finding the electron at a certain point. We define the square of the wave function, $(\psi.\psi^*.d\tau)$, as the probability of finding the particle in a volume element $(d\tau)$, located at a distance (r) at a certain time (t), and sidetrack the question as to what the wave actually is.

The electron's wave function, a mathematical entity, according to quantum theory, captures everything about the electron that we could possibly know. For example, the wave function of an electron confined to a spherical cavity is the three-dimensional description of how the electron's location is 'smeared out' over the space. In its lowest energy state, the wave function is spherical. The next highest energy level gives the wave function a dumb-bell shape.

Thus, quantum mechanics proposes a purely mathematical model for an atom, consisting of complex wave equations for each electron in an atom. Theoretically, we can find the wave function $\psi(r, t)$ that solves a particular problem in quantum mechanics, using Schrödinger's equation. When there is no force acting upon a particle, its potential energy is zero. We know the solution to this 'free' particle as a wave packet initially looks just like the Gaussian bell curve.

Unfortunately, one cannot always solve exactly the Schrödinger equation for a particular problem. We have to make certain assumptions to obtain an approximate solution for a particular problem. Wave packets, therefore, can provide a useful way to find approximate solutions to problems that one could not easily solve otherwise. First, we assume a wave packet to initially describe the particle under study. Then, when the particle encounters a force and its potential energy is no longer zero, this force modifies the wave packet. The problem is to find accurate and fast ways to 'propagate' the wave packet so that it still represents the particle at a later point in time.

The *four* quantum numbers (approximate solutions of quantum equation for an electron) help to physically describe the three-dimensional probability regions, known as the atomic orbital. The *first* quantum number, called the principal quantum number N, also referred to as the Shell "N" (maximum # of electrons in N shell equals $2N^2$), represents the radial extension of the region extending from the center of the atom. The *second* quantum number, called the subsidiary number L, represents the shape of the orbital.

The *third* quantum number, called the magnetic quantum number M, represents the direction in space that the orbital is pointing. The *fourth* quantum number, called the spin quantum number S, is part of the total angular momentum of a particle, atom, nucleus, etc. which is distinct from its orbital angular momentum. It represents the clockwise or counter-clockwise spin of an electron particle on an imaginary axis, or in terms of an electron manifesting itself as a field generating a wave disturbance. A particular elementary particle has a particular spin, just as it has a particular charge or mass.

The evolution of the concept of spin also has an interesting history. In 1922, Stern and Gerlach performed an experiment whose results they could not explain in terms of classical physics. Their experiment indicated that atomic particles possess an intrinsic angular momentum, or 'spin', and that this spin is quantized, i.e., it can only have certain discrete values.

The concept of spin is completely a quantum mechanical property, and cannot be explained in any way by classical physics. It is also important to realize that the spin of an atomic particle is not a measure of how it is spinning. In fact, it is impossible to tell whether something as small as an electron is spinning at all. The word 'spin' is just a convenient way of talking about the intrinsic angular momentum of a particle. The spin tells us how the particle would look from different directions.

According to quantum theory, spin is quantized, and it is restricted to multiples of $\hbar$ (= h/2 π, where h is Planck's constant. We characterize spin by a quantum number, S. For example, for an electron, S=+/-½, implies a spin of $+\hbar/2$ (= +h/4 π) when it is spinning in one direction and $-\hbar/2$ (= -h/4 π), when it is spinning in the other. Because of their spin, particles also have their own intrinsic magnetic moments, and in a magnetic field, the spin of the particles lines up at an angle to the direction of the field, precessing around this direction.

As stated, quantum mechanics limit our predictive capability. We can only assert that a particular outcome will occur with a certain probability. In other words, one can describe this indecision by saying that there are two alternative universes: one with an excited atom, one with a decayed atom. According to quantum mechanics, an atomic state will generally involve both universes coexisting and overlapping in a sort of hybrid reality.

However, the process of observation collapses the co-existing states into a single outcome that we observe. Before we go further in this direction and talk about the inherent uncertainty in quantum mechanics, let us briefly describe just one application of quantum mechanics - superconductivity. It has been worth over half a dozen Noble Prizes in physics. We also visit a newly emerging field, called nanoscience.

- Superconductivity

Superconductors are materials that offer no resistance to the flow of electricity. In 1911, Dutch physicist Onnes cooled mercury to the temperature of liquid helium at 4 degrees above the Absolute temperature, i.e. $4^\circ K$ ($T_c = 0^\circ K \sim -273^\circ F$). To his surprise, he found that it lost all of its electrical resistance. He received the Nobel Prize in 1913 for his research in this area.

In 1933, Meissner and Ochsenfeld discovered that a superconductor repels a magnetic field, a phenomenon known as diamagnetism, or the Meissner effect. The Meissner effect is very strong and it can levitate a magnet over a superconducting material. In 1935, Fritz London pointed out that superconductivity is a quantum phenomenon on a macroscopic scale, and diamagnetism is a fundamental property.

In 1962, Josephson discovered the tunneling phenomenon, known as the Josephson Effect. Josephson shared the 1973 Nobel Prize for this work. As a graduate student at Cambridge University, he had predicted that electrical currents would flow between two superconducting materials separated by an ordinary conductor or insulator. The quantum theory of normal metals was developed, but it was only in 1957 that Bardeen, Cooper, and Schrieffer advanced a theory of superconductivity, called BCS theory.

This theory could explain the phenomenon of superconductivity at temperatures close to absolute zero. Professor Bardeen was from my Alma Mater, University of Illinois, where I had the honor of meeting him in 1960. This was his second Nobel Prize for this work in 1972, having won one earlier for the transistor.

According to BCS theory, superconductivity occurs when electrons pair up and gather in single collective quantum states. In low-temperature superconductors, this crucial interaction among the electrons is due to the vibrations of the metal lattice (called phonons) of positive ions. This theory, however, cannot satisfactorily explain the phenomenon of superconductivity observed at higher temperatures.

After Onnes' discovery, scientists discovered other superconducting metals, alloys, and compounds. For example, Niobium-Nitride at 16oK was discovered in 1941, Vanadium-Silicon at $17.5^\circ K$ in 1953, the first commercial super-conducting wire of Niobium and Titanium alloy at

Westinghouse in 1962, and organic material superconductors in 1980. Collective, rather than individual, behavior of electrons seems to govern high-temperature superconductors and materials with giant and colossal magneto-resistance, but we do not know how.

Two IBM researchers, Muller and Bednorz (Nobel Prize in 1987), discovered in 1986 a brittle ceramic compound (a Lanthanum, Barium, Copper, and Oxygen compound - cuprates) that became a superconductor at $30^{\circ}K$. It triggered a flurry of activity as scientists began testing other ceramic (cuprate) combinations. As confirmed in 1994, a Thalium-doped mercuric cuprate exhibited superconductivity at $138^{\circ}K$. There have been results reported recently that certain materials become superconductors at temperatures as high as $250^{\circ}K$. However, we have a long way to go before we develop superconducting materials at room temperature. The most frustrating aspect is that we do not yet have the quantum theory to explain the precise workings of some superconductor materials, such as ceramics.

- Nanoscience

Emulating nature, scientists are trying to rearrange matter atom-by-atom, and study the resulting properties in a newly emerging field of science - nanoscience. This is similar to the way nature has built material and living objects, including nanostructures like a virus. In fact, biology is nanoscience. The DNA chain is a nanometer across. Science is ready to focus and zoom in on matters on a nanoscale, ($\sim10^{-9}m$), after studying it at the macro- and micro-scales. Scientists are discovering some startling properties of materials when they arrange atoms at nanoscale. In nanoscience, the quantum effects in the nanometer range drastically affect the properties of matter.

According to Feynman, the principles of physics do not rule out the possibility of maneuvering things atom by atom. It does not violate any laws, and it is doable. In fact, he offered a prize of $1,000 to the first person who could demonstrate nanotechnology. He wrote in an article that there is plenty of room at the bottom, and there is nothing in the laws of quantum mechanics that forbid machines the size of molecules. Recent developments in the field of nanoscience are truly astounding. For example, carbon nanotubes formed by arranging carbon atoms in hexagonal form can provide amazing properties. Such a nanotube can be 10,000 times thinner than a human hair. It is as strong as diamond, and it can bend or twist like steel. It can conduct heat and electricity, or act as a semiconductor.

Within a decade, nanotechnology, based on nanoscience, has the potential to usher in a revolution. The National Science Foundation predicts that, by 2015, nanotechnology could easily grow into a $1trillion annual industry. This revolution would be in the areas of consumer electronics through miniaturization, and through nanotechnology in chemicals and basic materials. In energy, it would be through thin-film photoelectric devices and fuel cells.

In pharmaceutical and medical technology, it would be through sensing, diagnostic, and therapeutic nanomedicine, with nanocapsules administering medicine to individual cells. Nanodevices have already successfully located tumors just under the skin of laboratory rodents. Polymersomes about 10 nm, injected into the body, could seek out and bind specific types of cancer cells.

F. Uncertainty Principle

Continuing our journey through quantum mechanics, we have already noticed that quantum mechanics has dealt a severe blow to the deterministic world of science by introducing probabilities. Heisenberg added another element to this debate in 1927 when he stated his uncertainty principle. Uncertainty, as shown by Heisenberg, is the basis of quantum mechanics

Heisenberg challenged the notion of the simple deterministic model in terms of causality in nature, which states that a predictable effect follows every determinate cause in nature. In classical physics, this meant that one could determine the future motion of a particle from the knowledge of its present position, its momentum, and all of the forces acting upon it. The uncertainty principle implies that one cannot know the precise position and momentum of a particle at a given instant. Its future, therefore, cannot be determined.

Heisenberg's Uncertainty Principle turned out to be an essential component of the broader interpretation of quantum mechanics, the Copenhagen interpretation. It reads,

"We regard quantum mechanics as a complete theory for which the fundamental physical and mathematical hypotheses are no longer susceptible of modification."

It is interesting to note how famous scientists can strongly disagree with each other. Schrödinger, Einstein, and Bohr did not agree with Heisenberg.

In fact, Heisenberg's battle with Bohr grew so intense in the early months of 1927 that Heisenberg reportedly burst into tears at one point. However, many scientists now believe that the implications of the uncertainty principle for physics, philosophy, and the perception of reality, are likely to be more far reaching than Einstein's theories of relativity. Elimination of the idea of strict causality and the notion of particles-antiparticles popping out of quantum space, and annihilating each other, are some of the overwhelming implications of this principle.

What is the basis of Heisenberg's Uncertainty Principle? A measurement process involves information, or energy transfer, between the observed bodies and an observer through the experimental apparatus. Thus, any measurement process interferes with the experiment itself. In classical mechanics, the perturbations and uncertainties of the measurement process are insignificant. Thus, we can measure simultaneously any quantity with the required precision. However, in quantum physics, the quantities measured are so weak that the perturbations of the measurement process become significant. This changes our perspective radically.

Heisenberg was the first to realize that certain pairs of measurements have an intrinsic uncertainty associated with them. Heisenberg's uncertainty relations link together the uncertainty in the measurement of position (Δx), momentum (Δp), energy (ΔE), and duration (Δt). Heisenberg's Uncertainty Principle states that the process of measuring the position (x) of a particle disturbs the particle's momentum (p). Thus, $\Delta x \, \Delta p \geq \hbar \, (= h/2\pi)$, where Δx is the uncertainty of the position and Δp is the uncertainty of the momentum, and h is Planck's constant. Also, $\Delta E \, \Delta t \geq \hbar \, (= h/2\pi)$. The derivation of this second relation has been quite problematic.

It tells us essentially that we can never know both the position and the velocity of an elementary particle at any moment, since the product of

uncertainties in the measurement of the quantities has to be at least
For instance, if we know exactly where something is located, then we
cannot know with certainty as to how fast, or in what direction, it is
moving. Thus, one cannot calculate the precise future motion of a
particle, but can calculate only a range of possibilities for the future
motion.

One could calculate exactly the probabilities of each motion and the
distribution of particles following these motions from Schrödinger's
wave equation. The uncertainty principle does not imply that everything
is uncertain. Rather, it tells us exactly where the limits of uncertainty lie
when we make measurements of sub-atomic events.

We do not notice this in everyday life, because any inherent
uncertainty is well within the acceptable accuracy we desire. For
example, we may see a person standing and think that we know exactly
where he is. However, do we really know exactly? To test it, we might
decide to measure the position of the person to an accuracy of a trillionth
of a centimeter. When we attempt to do so, we would be trying to
measure the position of the individual atoms that make up the person.
These atoms would be moving around, as long as the temperature of the
person is above absolute zero degrees Kelvin!

Thus, the uncertainty principle describes the behavior of matter at the
atomic level. Unlike classical physics, it interjects probabilities, and
forces us to observe reality in a radically different way. Given a
particular atomic state, we cannot generally predict how it will change.

The question arises whether it is the measurement difficulty alone that
prevents us from knowing, with total precision, both the particle's
location and velocity. In such a case, Einstein argued that an electron
still has an exact location, velocity, and the quantum reasoning merely
points out our limitation to measure both of them exactly. However, as
shown by scientists Bell, Alain and others, it is not just a measurement
limitation. We simply cannot describe electrons as simultaneously
having an exact location and velocity.

- Uncertainty Principle & the Planck Scale
The uncertainty principle also implies that the energy and momentum
in a microscopic region of empty space are uncertain. If we try to

confine a particle within an extremely small box, it becomes very frantic, and its speed and momentum increases. Heisenberg's Uncertainty Principle indicates that the energy and momentum fluctuate violently and shift back and forth.

As shown by Dirac, this violent exchange in the empty region of space can result in the momentary creation of an electron and its antimatter particle positron. These particles are supposed to annihilate each other immediately, giving back the borrowed energy. In fact, this seething frenzy of the space fabric at Planck scale is directly responsible for the conflict between relativity and quantum mechanics.

This frantic activity is particularly important near the Planck's scale (Table 1.2)

Planck Scale	Symbol/Unit	Value	Expression
Planck Mass	m_p kg	2.1767×10^{-08}	$\sqrt{(\hbar.c/G)}$
Planck Length	l_p m.	1.6161×10^{-35}	$\sqrt{(\hbar.G/c^3)}$
Planck Time	t_p sec.	5.3906×10^{-44}	$\sqrt{(\hbar.G/c^5)}$

Constants	Symbol		Units
Planck's Constant	$\hbar = h/2\pi$	1.0545×10^{-34}	$kg.m^2/sec$
Gravitational Constant	G	6.6726×10^{-11}	$m^3/(kg.sec^2)$
Speed of light	c	$2.9979 \times 10^{+08}$	m/sec

Table 1.2 – Planck Scale

The Planck length is the scale at which classical ideas about gravity and space-time cease to be valid, and the quantum effects dominate. This is the 'quantum of length', the smallest measurement of length with any meaning. It is roughly about 10^{-20} times the size of a proton ($\sim 2.1 \times 10^{-16}$ m). The Planck time is the time that a photon takes to cross the Planck length, traveling at the speed of light. This is the 'quantum of time', the smallest measurement of time with any meaning.

Within the framework of the current laws of physics, we can talk about the evolution of the Universe only after it was $\sim 10^{-43}$ sec old. The Planck length arises naturally, when we consider the ultimate limits to measurement. Roughly speaking, it happens because a finer length measurement requires large momentum. To determine the position of a particle, we shine it with a wave, and we cannot determine the position of a particle more accurately than the wavelength (distance between the wave crests).

For example, to measure the position of a particle with an uncertainty of Planck length scale, $l_p = 1.616 \times 10^{-35}$ m, we would need to shine it with a wave of wavelength l_p. The corresponding frequency would be $1.8551 \times 10^{+43}$ Hz. Since energy comes in quanta, we have to use at least one quantum of energy at this short wavelength, which according to the relation, E = h.f. equals $1.221 \times 10^{+19}$ GeV. This huge amount of energy would impart a very high momentum and disturb the speed of the particle, making the speed measurement more uncertain. Thus, the more accurately we try to measure a particle's position, the less accurate its speed measurement becomes.

When the momentum becomes this large, the gravitational effect becomes strong. This curves space-time, and distorts the interval that one seeks to measure. Thus, a fundamental difficulty arises in resolving lengths below the Planck scale. This point has been 'rediscovered' many times. The uncertainty principle, which is central to quantum theory, dictates that probing on a smaller scale requires higher energy or, equivalently, extremely small wavelengths.

The tiny Planck scale corresponds to a massive $1.22 \times 10^{+19}$ GeV of energy, implying an extremely high energy density within a quantum of space. Such energy might have been available to particles only during the first split second of the Big Bang. It is almost 100 trillion times that of the highest energies achieved in today's particle accelerators.

One might think that the Planck scale is more fundamental than the other scales. However, as Wilczek points out, the quantities one chooses to regard as fundamental depend on what domain one seeks to describe. For example, much of chemistry and molecular biology can be described by taking only the electron mass and charge as inputs, using Planck's constant $\hbar = h/2\pi$ as the unit of action, and regarding the atomic nuclei as infinitely massive point-particles.

In this system, the Bohr radius, $\hbar^2/e^2 m$, appears as the fundamental unit of length, which sets the scale for atomic and molecular sizes. Similarly, one can describe strong interaction physics by taking only the quantum chromo dynamics mass scale as input, using Planck's constant and the speed of light (c) as units of action and velocity. In this system, the fundamental unit of length is $\Lambda/\hbar c$, which sets the scale for proton and nuclear sizes.

G. Four Force Fields

According to our current understanding of the Universe, one could explain its diversity in terms of subatomic particles and the following four fundamental interacting forces.

Gravity - Gravitational force acts between matter particles. It is the weakest force but has infinite range, and every particle in the Universe feels its influence. We can ignore gravity on the atomic scale because it is so weak. However, its influence increases with every atom added to an object, and it becomes significant once an object has enough atoms to be visible.

Electromagnetic – Electromagnetic force interacts with electrically charged particles but not electrically neutral particles. It is stronger than gravity ($\sim 10^{38}$ times stronger than gravity), and it too has infinite range, but it affects only charged particles. It holds atoms and molecules together. In practice, the electromagnetic force has a limited range because of the matching north and south magnetic poles, and the balance between oppositely charged particles.

Strong - Strong nuclear force holds quarks together in protons and neutrons in the nucleus. It is stronger than all the forces ($\sim 10^{41}$ times stronger than gravity). It has a very short range ($\sim 10^{-15}$m) and exerts no influence outside an atomic nucleus. It holds the protons and neutrons together within the nucleus, and it is over a hundred times stronger than the electric force between the protons.

Weak - Weak nuclear force, responsible for radioactivity, acts within the nucleus on all matter particles having an odd spin. It is weak, but stronger than gravity ($\sim 10^{27}$ times stronger than gravity). It has a very short range ($\sim 10^{-15}$m). It is responsible for splitting, in about 15 minutes, a neutron into an electron and antineutrino.

The existence of these different force fields is responsible for the life we know. For example, the Sun's fuel comes from the nuclear reactions based on weak force, but gravity holds the Sun together. If the weak force were a little stronger than gravity, the Sun would have burned out long ago. If it were a little weaker, it would provide less energy. In either case, there would be no life on Earth.

H. Single Unified Field Theory

The interaction strengths of the four known fields depend on the energy at which they are measured. Experiments suggest that, at high enough energies (10^{16}GeV), electromagnetic and weak forces become part of a single electroweak force, as shown in Fig. 1.10.

.

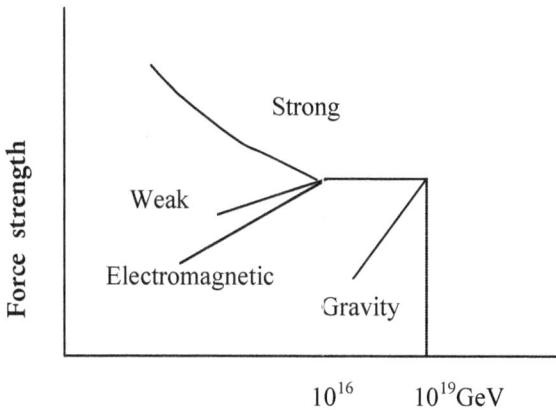

Fig. 1.10 Four Fields

This split occurred due to events in the very early history of the Big Bang. However, one cannot trace this early history without a better theory of gravity and other forces, which split at lower energies. It is interesting to note that the extrapolation of the interactions of fields shows that all the fields, except gravity, become equal at around 10^{16} GeV. The gravitational field also has the same strength around 10^{19} GeV, as shown in Fig. 1.10

We can unify strong forces with the electroweak forces, but the inclusion of gravity presents a serious challenge. The initial idea behind the development of the Grand Unified Theory (GUT) was to unify all three fields. As stated, Einstein in his later years was obsessed with the idea of finding a unified field theory. He did not succeed. Then, we did not know all the four fields, and Einstein did not believe in quantum mechanics or the uncertainty principle. Many scientists now believe that the uncertainty principle might be a fundamental component of the unified theory.

What do we expect from the unified field theory? Unified theory is a mathematical framework in which different fields and particles occur naturally. Such a unified theory, according to current thinking, might also answer questions about the true nature of mass. In such a theory, we should not have to assume the mass and charge for different particles from experiment, as we do in the Standard Model. At present, the Standard Model has no mechanism to account for the masses of the particles, unless we supplement it with 'scalar' fields. The interaction of the other fields with scalar fields can give masses to the particles of the Standard Model. A unified theory should be able to predict these values.

Obviously, it is an ambitious goal. To discover such a unified theory, we must look for mathematical models having considerable elegance and consistency. If we find a candidate for such a theory, we first need to understand the detailed structure of the theory. The next step would be to check whether it describes nature. We must also keep in mind certain guidelines when constructing a general mathematical framework for a unified theory. We need to insure that all the physical phenomena are independent of the coordinate system chosen to map out time and space.

We visit the symmetry in nature next, when we discuss the Standard Model. Symmetry violation plays an important role in the unified field theory. Some scientists believe that the study of symmetry violations might lead to the ultimate unified theory, which would unify all the known forces and particles. They believe that the relativity postulates are mere approximations of a more general theory. However, from the success of the relativity theories, quantum mechanics, and the Standard Model, it would seem that the Lorentz and CPT symmetry violations, if any in the unified theory, would have to be extremely small.

H. Standard Model of Particle Physics

We are now ready to visit the currently accepted Standard Model of particle physics. It is not easy to comprehend all aspects of the Standard Model. The Standard Model describes all ordinary matter and all the force fields except gravity, based on quantum field theory. The exclusion of gravity implies that the model, while impressive, is clearly not the final theory. It does explain hundreds of particles and complex interactions in terms of a few fundamental particles and interactions.

Science has penetrated the nucleus of an atom to show that protons and neutrons are not fundamental particles. There are more fundamental particles, leptons and quarks. According to the Standard Model, there are six types of quarks and six types of leptons. Table 1.3 lists the particles in the Standard Model and their properties.

The two least massive types of quarks and the least massive charged lepton, the electron, make up almost all the stable matter of the Universe..Weak interactions are responsible for this. As this field decays, more massive quarks and leptons produce lighter quarks and leptons When a particle decays, it disappears and, in its place, two or more particles appear. The sum of the masses of the produced particles is always less than the mass of the original particle. This is why stable matter around us has only the lightest particles, electron, and two quarks (up and down). When a quark or lepton changes type, i.e. a muon changing to an electron, its 'flavor' also changes. All flavor changes are due to the weak interaction.

We assign various properties to different particles. Fermions are the fundamental matter particles, named after the famous scientist Fermi. A fermion has an odd half-integer (1/2, 3/2, ...) intrinsic angular momentum (spin), measured in units of ℏ (= h/2 π == 1.055 x 10^{-34} J s). Essentially, fermions include two types of fundamental particles: leptons and quarks. An electron is a lepton, and quarks are constituents of protons and neutrons. Quarks cannot exist alone. They only exist with other quarks, and form composite particles called hadrons.

There are two classes of hadrons:

1) Baryons - Baryons are any hadron made of three quarks (qqq). For example, protons are two up quarks and one down quark (uud), and neutrons are one up and two down quarks (udd).

2) Mesons - Mesons contain one quark and one antiquark.

Although individual quarks have fractional electric charges, one cannot directly observe these fractional charges. The sum of the quarks' electric charges in a hadron is always an integer number. While individual quarks carry color charges, hadrons are color-neutral - a property called confinement, which rules out observing a single quark.

Leptons

Flavor	Mass(GeV/c²)	Elect. Charge
u up	.005	+2/3
d down	.01	-1/3
c charm	1.5	+2/3
s strange	0.2	-1/3
t top	180	+2/3
b bottom	4.7	-1/3

Quarks

	Flavor	Mass (GeV/c²)	Elect. Charge
ν_e	e neutrino	$< 7 \times 10^{-9}$	0
e^-	electron	.000511	-1
ν_μ	μ neutrino	< .0003	0
μ^-	muon	0.106	-1
ν_τ	τ neutrino	< .03	0
τ^-	tau	1.7771	-1

Table 1.3 – Standard Model, Leptons and Quarks Properties

Bosons are those particles that have an integer spin measured in the units of ℏ (spin = 0, 1, 2...). The following are bosons: carrier particles related to all fundamental interactions, to be discussed soon, and the composite particles, with the even numbers of fermion constituents (such as mesons). The nucleus of an atom is a fermion or boson, depending on whether the sum of the number of protons and neutrons is odd or even. For example, the nuclei of helium bosons - and thus, very cold helium, a superfluid (no viscosity, etc.) - has strange behavior, since bosons in the nuclei may pass through each other.

Quarks and gluons (glue quarks together) have a type of charge that is not quite electromagnetic - rather, it is a "color" charge. The force between color-charged particles is very strong, hence the name 'strong' force. This force holds quarks together to form hadrons. Therefore, we call its carrier particles 'gluons', since they successfully 'glue' the quarks together. Note that hadrons (such as protons and neutrons) are color-neutral, as are leptons. Thus, the strong force only acts on the small level of quark interactions.

All particles have to follow certain rules. For example, physicists use the Pauli Exclusion Principle to categorize the fundamental particles into two classes: fermions that are subject to the Pauli Exclusion Principle, and bosons, which are not subject to the Pauli principle. Pauli Wolfgang discovered the principle in 1925, for which he received Nobel Prize in 1945. It states that particles of half-integer spin (fermions) must have anti-symmetric wave functions, and particles of integer spin (bosons) must have symmetric wave functions. In addition, no two fermions can exist in the same place at the same time. Because of this odd half-integer angular momentum, all fermions obey Pauli's Exclusion Principle - a very important property of ordinary matter.

The Standard Model gives very precise results, once we follow certain rules and put in it the experimentally measured values for 19 parameters. These parameters include mass of various particles and the interaction forces. The Standard Model faces two main challenges. Firstly, one cannot directly calculate the 19 parameters. Secondly, it cannot describe gravitational interaction. Unfortunately, scientists have not been able to discover all the real physics behind the Standard Model.

The Standard Model evolved as follows. The inclusion of special relativity into the Schrödinger equation, by Klein and Gordon, led to the development of quantum field theory, which, along with the understanding of strong and weak fields in the atom, led to the Standard Model for the atom. Thus, scientists developed the Standard Model with the help of quantum field theory.

In fact, the Standard Model is quantum field theory, which is an application of quantum mechanics to fields. Quantum field theory owes its origin to the problem of computing the power radiated by an atom when it dropped from higher energy quantum state to a lower energy state. Quantum field theory followed Dirac extension of Schrodinger's equation to include special relativity, and it involved quantum treatment of the electromagnetic field. The Standard Model includes three fields, but not gravity. Little ripples, known as quanta, in these fields carry energy from one point to another. We recognize these ripples in the laboratory as elementary carrier particles, e.g. photons for the electromagnetic field.

- Standard Model & Symmetry

Feynman defines symmetry in simple terms, saying that anything is symmetrical if one can subject it to a certain operation, and it appears the same after the operation. The symmetry of space and time played an essential role in the development of relativity, and it does so in the development of the Standard Model. Nature observes symmetry in time and space, since it does not discriminate between one moment and location from the other. The physical laws do not depend on where and when we use them. In relativity theories, we observed that all physical laws are the same in all reference frames, and we treat all observers identically.

According to Wigner, the symmetry in space and time plays a critical role in discovering the laws of nature. Wigner, in his book, Symmetries and Reflections, says that if the correlation between events changed from day to day and from place to place, it would be impossible to discover them. Wigner also thinks that the concept of symmetry principles essentially relates to our ignorance. In other words, if we could directly know all the laws of nature, we would not need to use symmetry principles in our search for them.

The laws of physics also honor rotational symmetry in that they do not change, just because you observe them from a different angle. In order to maintain various symmetries in the Standard Model, certain force fields were required. Certain symmetries - called gauge symmetry – relate to the strong and weak fields in an atom. Thus, consideration of various types of symmetries played an important role in the development of the Standard Model.

We now visit Lorentz symmetry, which is the basic building block of both general relativity theory and quantum field theory, describing all observed phenomena. We can describe Lorentz symmetry in terms of Lorentz transformations. Lorentz transformations are of two basic types: rotations and boosts. There are three possible rotations, one about each of the three spatial directions. A boost is a change of velocity, and there are three possible boosts, one along each of the three spatial directions.

A physical system is said to have 'Lorentz symmetry' if the relevant laws of physics are unaffected by Lorentz transformations (rotations and boosts). Since the general Lorentz transformation involves both of these at the same time, it leads to the mixing of space and time intervals. Lorentz transformations between different co-ordinate frames ensure that the space-time interval and the value of the speed of light are the same for all inertial observers.

Let us now discuss some other types of symmetries: charge conjugation (C), parity inversion (P), time reversal (T), and their combinations. C-symmetry converts a particle into its antiparticle. P-symmetry transforms an object into its mirror image, turned upside down. T-symmetry changes the direction of the flow of time. C-symmetry implies that the laws of physics are same for particles and antiparticles. P-symmetry implies that the laws are also the same for mirror image particles (particles spinning in opposite directions). T-symmetry implies the laws to be the same for forward and reverse directions of time.

Lee and Yang found that the weak field did not obey P-symmetry, nor does it obey C-symmetry. However, it obeys combined CP-symmetry. In general, the Universe does not obey CP-symmetry. In other words, if one replaced the particles by antiparticles and they were mirror image particles but did not reverse time, then the laws of physics would have to be different. These symmetry considerations have far-reaching consequences for matter and force particles in the Standard Model, and in the evolution of the Universe. For example, it explains why quarks, and not anti-quarks, constitute all matter.

Combining three transformations forms the CPT transformation: charge conjugation (C), parity inversion (P), and time reversal (T). A system is said to have 'CPT symmetry' if the physics is unaffected by the combined transformation, namely, the CPT transformation. Combined CPT symmetry has to hold if Lorentz symmetry is obeyed. Therefore, all systems that obey quantum mechanics and relativity obey combined CPT symmetry. These symmetries are the basis for Einstein's relativity. Experiments show, to exceptionally high precision, that all the basic laws of nature seem to have both Lorentz and CPT symmetry.

Extending the notion of symmetry to spin, which denotes the angular momentum of a particle as it spins around, results in the definition of supersymmetry. Supersymmetry predicts a partner for each fundamental particle, whose spins differ by a half unit, i.e. its intrinsic angular momentum differs by $\hbar/2$. For example, spin-0 superpartner of electron is selectron; having a ½ spin, a neutrino's superpartner is sneutrino, and a quark's super partner is squark. Similarly, all force carrier particles have the corresponding spin-1/2 superpartners, photino for photon, gluinos for gluon, and winos and zinos for W and Z bosons, respectively.

A candidate particle for dark matter, neutralino, is perhaps an amalgam of photini, zino and perhaps other partners predicted by supersymmetry. We have not observed these new partner particles, perhaps because they are massive compared to the known particles and beyond the reach of current particle accelerators.

Supersymmetry is appealing, not just because we expect nature to observe spin symmetry, but also because it solves a basic problem in the Standard Model. As mentioned in Heisenberg's Uncertainty Principle, space, when examined at the microscopic level, is in quantum mechanics frenzy with violent momentum and energy fluctuations. In view of this, numerical parameters put in the Standard Model require accuracy of one part in a million billion to cancel out quantum effects in certain processes involving particle interactions. This is too delicate a fit, but supersymmetry can resolve this problem.

Supersymmetry insures that each boson and fermion occur in pairs, resulting in an equal number of them. Now, fermions with half-integer spin have negative ground state energy and bosons with integer spin have positive ground state energy due to quantum fluctuations. Thus, the ground state energies cancel out at the outset in this supergravity scheme. We do not have to rely on the extremely close accuracy required for certain parameters in the Standard Model.

- Standard Model and Symmetry Violation

Some scientists are also looking for the spontaneous violation of symmetry. Such violations could play a critical role in the extension of the Standard Model, in the development of unified field theory, in the development of quantum gravity and in making quantum mechanics compatible with relativity theories. Several quantum gravity theories, loop quantum gravity, string theory, and non-commutative geometry involve Lorentz violations in some regime. Lorentz symmetry violation is somewhat similar to the breaking of other kinds of symmetries, when the physical laws are symmetric but the actual system is not.

Kostelecky, in the September 2005 issue of the Scientific American, gives the example of a thin rod placed vertically with one end on the floor and the other end subjected to a downwards vertically applied force. The rod has symmetry under rotation, but if we keep increasing the applied force, the rod will eventually bend in some direction, breaking the rotational symmetry.

Kostelecky looks at Lorentz symmetry in two ways. The observer Lorentz invariance transformation implies that the laws of physics are the same for all inertial observers. In other words, the laws of nature cannot depend on the perspective of an observer. For example, a person on a moving train and a person waiting at a station will obey the same laws of physics. The particle Lorentz transformation, on the other hand, concerns a particle or an object in motion with respect to a fixed inertial frame. It answers the question, *'What would happen if the person on the train starts walking inside the train?'* In other words, from the fixed perspective of a moving train, do a seated passenger and a moving passenger experience the same laws of physics?

Both the particle Lorentz transformation and the observer Lorentz transformation are the same in the absence of Lorentz violation. For example, the moving passenger would introduce a third observer frame, but the laws remain the same. However, Kostelecky shows that these two types of transformation are not equivalent for Lorentz symmetry violation. In other words, the laws of physics experienced by the moving passenger can be different to those felt by the passenger who remains seated.

As shown by Kostelecky and Samuel, string theory permits Lorentz symmetry to be spontaneously broken in the early Universe. This would imply that small relic background fields would permeate the Universe, and point in spontaneously chosen directions. In the presence of one of these relic fields, an elementary particle would experience interactions with a preferred direction in space-time. In other words, there could be preferred directions in 3D space in any fixed reference frame. Of course, all interactions would remain invariant under observer Lorentz transformations, but the particle Lorentz invariance would be broken.

Scientists suspect that Lorentz symmetry can break down by some mechanism originating at the Planck scale. Such spontaneous Lorenz symmetry violations could lead to the Standard Model Extension. This would lead to the true unified field theory. Such an extension would contain all possible generalizations of relativity, compatible with the Standard Model and gravity.

Colladay, Kostelecky and Potting did develop such a theory, called the Standard Model Extension (SME). The Standard Model Extension (SME) describes all the particle interactions that maintain observer Lorentz invariance but not particle Lorentz invariance. At present, nothing in the current Standard Model of particle physics permits the violation of special relativity.

Spontaneous Symmetry breaking also has other consequences for quantum field theory. For example, according to a theorem first stated by Goldstone in 1960, in the case of a global continuous symmetry, mass-less boson (known as "Goldstone bosons") appear with the spontaneous breakdown of the symmetry. First seen as a serious problem, it became the basis for the solution of another similar problem -- by means of the so-called Higgs mechanism. In 1954, the Yang-Mills theory of non-Abelian gauge fields predicted unobservable mass-less particles, the gauge bosons.

According to a 'mechanism' due to Higgs et al, when the internal symmetry is promoted to a local one, the Goldstone bosons 'disappear' and the gauge bosons acquire a mass. This mechanism for the mass generation for the gauge fields ensures the renormalizability of theories involving massive gauge fields, such as the Glashow-Weinberg-Salam electroweak theory, developed in the second half of the 1960s. In dynamical symmetry breaking theories, as the unified model of electroweak interactions, the spontaneous symmetry breaking is responsible (via the Higgs mechanism) for the masses of the gauge vector bosons. This is due to the symmetry-violating vacuum expectation value of scalar fields, called 'Higgs fields'.

Scientists have not experimentally observed the scalar fields or Higgs particles. They have designed a number of experiments to obtain evidence for relativity and symmetry violations, but thus far, the results are negative. Some scientists, especially the string theorists, now believe that we need Planck scale sensitivities to detect such violations.

- Standard Model & Dark Matter

As if the complexity, constituents, and the interacting forces of matter did not already baffle us, physicists now tell us that dark matter, which we cannot see, constitutes most of the matter in this Universe. This prediction is because the ordinary matter we see in the Universe is not enough to hold the Universe together. Without some additional matter, galaxies would not have enough gravitational pull to hold themselves together with just the known matter in the Universe.

A team of US astronomers announced recently at the American Astronomical Society meeting in Washington that the giant halo of dark matter that surrounds our galaxy is shaped like a flattened beach ball. They inferred the shape of the 'dark matter halo' from the path of debris left behind, as the Sagittarius dwarf galaxy slowly orbits the Milky Way.

To explain the existence of dark matter, Milgom, in the August 2002 issue of the Scientific American, suggested that we modify Newton's second law, relating acceleration to the force at extremely low acceleration below ~10^{-10} meter per sec per sec (e.g. acceleration of our solar system toward the center of our galaxy). At such a small acceleration, his MOND model proposes a force proportional to the square of acceleration, which accounts for the discrepancy instead of assuming the existence of dark matter. MOND also has some problems, which according to Milgrom, could be either unimportant or fatal.

Dark matter, as its name suggests, does not reflect light, and rarely, if ever, interacts with visible matter. In fact, the only evidence that it exists is its gravitational pull on light and stars. Dr. Chung-Pei Ma of the University of California, Berkeley said,.

"Dark matter leaves its imprints everywhere, but we still don't know what it is."

As stated earlier, recent measurements by scientists with WMAP show that the contents of our Universe include 4% atoms (ordinary matter), 23% of an unknown type of dark matter, and 73% of a mysterious dark energy. We shall return to the discussion of dark energy.

Scientists are now vigorously searching for the particle that constitutes dark matter. At present, the leading candidate for dark matter comes from supersymmetry, which predicts heavy supersymmetric partners of electroweak gauge bosons and the Higgs field that are electrically neutral and do not react with the electromagnetic radiation.

For example, scientists think that nutralinos are fermionic partners of the neutral gauge bosons and Higgs field; they have large mass and interact weakly. There are several fancy names given to cold dark matter particles, such as neutralinos, weakly interacting massive particles (WIMPs), massive and compact halo objects (MACHOs).

The WIMP is supposed to have 50 to 100 times the mass of a proton. However, we have yet to detect it despite serious attempts. It is mainly because a WIMP tends to pass straight through ordinary matter. We have already talked about the neutralino being a promising candidate, since it is heavy and neutral in charge. Scientists are also looking for MACHOs in space.

MACHOs are believed to be ordinary matter comprised of protons, neutrons, and electrons. They are like white dwarfs, neutron stars, or even black holes that are not readily detectable. Scientists focus powerful cameras attached to telescopes, looking away from a galaxy, to study variations in the brightness of a star; these variations are due to the bending of light because of gravitational field variations when MACHOs pass near the star.

Scientists are looking for the controversial Higgs boson, believed to be quanta of the scalar field. The interaction of this scalar field with other fields of the Standard Model might give mass to the particles in the Standard Model. As stated, the scientists have not been able to discover the scalar field and its quanta, the Higgs particle. The scalar field, unlike the other fields in the Standard Model, does not carry a sense of direction and is supposed to pervade all space. Lederman has also called the Higgs particle the 'God particle'.

Scientists hope to discover the presence of the Higgs particle in Fermilab's Tevatron, a 7-mile-long circumference particle accelerator. This accelerator would smash opposing beams of protons and antiprotons around a circular track, sifting through the debris with two immense detectors called CDF and D0. The Higgs particle is supposed to play an important role in the Standard Model, and proving its existence or absence could change the entire foundation of physics. The results might indicate the existence of particles and forces not yet imagined and might call for an entirely new set of laws.

As stated, many scientists also believe that the Higgs boson is the key to unlocking the mystery of the elementary particles: the quarks and the leptons. The Standard Model does not give us the answers to many questions: Why are there three 'generations' of matter particles? Why do they have the masses and electric charges that they do? There is no good theory yet as to why different particles have different masses. Some scientists believe that the Higgs bosons are related to the mechanism by which the matter particles acquire mass. Scientists have very expectations from the discovery of the Higgs particle.

According to Dave Rainwater at Fermi Laboratory, the Higgs boson is interesting. We have this as the only reasonable explanation for the origin of mass. Without the Higgs boson, all fundamental particles would have no mass, and the Universe would be very different. The weak nuclear forces wouldn't be weak at all, for instance, so the elemental composition of the cosmos would be radically different; stars would shine differently, and we probably wouldn't exist. The co-spokesman of the experiment at Fermilab said

"One thing we expect the Higgs to open up is the question of supersymmetry. Supersymmetry is a relationship between the particles of matter and the forces of the Universe. Mathematically, it is beautiful. Not one piece of direct experimental data really supports it yet. Finding a Higgs in the place we expect would be a piece of evidence. Not finding it would be a big problem for the advocates of this idea. What would shake the foundation of physics much more than finding the Higgs would be a definitive 'ruling it out'. That would upset all of our conceptions about how the Universe works. It would make supersymmetry something that, if it applies in the Universe, does so only at much higher energies than we can observe. And it would require new forces or new laws to explain masses, in the absence of a Higgs."

- Standard Model & Antimatter

The Standard Model also helped predict the existence of antimatter. Dirac, while explaining the half-spin of an electron in 1928, also predicted electron's partner – antielectron or positron (electron-like but with positive electric charge). We now understand that all fermion particles have corresponding antiparticles, but photons and other carrier particles (bosons) are their own antiparticles.

Carl Anderson confirmed the existence of the positron in 1932, while studying cosmic showers in a cloud chamber. The discovery of the antiproton by Segre and Chamberlain soon followed in 1955, and the discovery of the antineutron by a team of scientists came soon afterwards. The antinucleus (antideutron), consisting of one antiproton and one antineutron, was formed in 1965, and 9 atoms antimatter (antihydrogen – composed of an antiproton and a positron) were produced at CERN in 1995 in a LEAR machine. There is intense ongoing research in this area.

A US team of scientists reported, in the 2007 journal Nature, the discovery of Di-positronium, a new molecule formed from an electron and a positron. In 1946, they predicted the existence of the molecule, but it had remained elusive to science until now. The US team has now created thousands of these molecules by merging electrons with their antimatter equivalent, positrons. It could be a key step in the creation of ultra powerful lasers, known as gamma-ray annihilation lasers.

Antimatter is indeed a revolutionary idea. Matter and antimatter can never exist together as they annihilate each other instantly, generating a great deal of energy. When a positron and an electron annihilate each other, their mass is converted into energy in the form of two gamma ray photons, emitted in opposite directions, each with the energy of 511 KeV. Particle physicists use positron-electron annihilations to study matter and fundamental forces at high energy, creating new particles and antiparticles in machines, such as the Large Electron-Positron Collider (LEP) at CERN.

Since for each basic particle of matter, there exists an anti-particle with the same mass but the opposite electric charge, processes occurring at the beginning of the Universe should have left us with equal amounts of matter and anti-matter. However, we live in a Universe made up overwhelmingly of matter. Why?

Recently, scientists found a clue to this paradox when, working on the DZero experiment observed collisions of protons and anti-protons in Fermilab's Tevatron particle accelerator. They found that these collisions produced pairs of matter particles slightly more often than they yielded anti-matter particles. The results show a 1% difference in the production of pairs of muon (matter) particles and pairs of anti-muons (anti-matter).

According to Stefan Soldner-Rembold, one of the spokespeople for Dzero, it was a very unusual observation. Physicists had already seen such differences - known as called "CP violation". But these known differences are much too small to explain why the Universe appears to prefer matter over anti-matter. These new findings, submitted for publication in the journal Physical Review, show much more significant "asymmetry" of matter and anti-matter - beyond what can be explained by the Standard Model.

At low energy, scientists have used electron-positron annihilations to reveal the workings of the brain in Positron Emission Tomography (PET scan). In PET, the positrons come from the radioactive nuclei, put in a special liquid injected into the patient. These emitted positrons then annihilate with the electrons in the nearby atoms in the brain, emitting gamma rays in two opposite directions, thus conserving momentum.

At present, it is quite time consuming and expensive to produce antimatter. For example, it takes almost a year to produce one billionth of a gram of antihydrogen, costing several hundred million dollars. Antimatter is electrically neutral, but it has magnetic properties and can be stored in magnetic bottles. There is speculation about using antimatter engines for propulsion, just as the Enterprise does in Star Trek, and to use annihilation energy for weapons. However, we shall have to wait for such developments, until scientists can produce enough antimatter and perform some experiments.

- Standard Model & Force-Carrier Particles

According to the current understanding gained from the Standard Model, each type of fundamental force is 'carried' by force-carrier particles, which are also called 'messenger' particles. The Standard Model proposes four interactions between particles. Table 1.4 summarizes the four fields and their carrier particles.

Force Field	Carrier Particle	Mass	Field Strength
Gravity	Graviton	0	Very Weak
Weak	W^+, W^-, Z^0	86, 97	10^{27}x Gravity
EM	Photon	0	10^{39}x Gravity
Strong	Gluon	0	10^{41}x Gravity

Table. 1.4 – Four fields and carrier particles

The photon, a carrier particle for electromagnetic fields, carries the message between opposite charges to attract and between similar charges to repel each other with a certain force. The carrier particles, called 'carrier bosons', have integer spin; they do not obey the Pauli exclusion principle, and can occupy the same location.

These forces, or interactions, which affect matter particles are due to the exchange of carrier particles: photons for the electromagnetic field; gluons for the strong field; W+, W- and Z bosons for the weak field; and graviton (suspected but not observed) for gravity. We have visited gravity earlier. Its carrier particle, graviton, has not yet been observed. Gravity does not fit in the Standard Model.

Recent theories concerning the electromagnetic field suggest that it is a quantum field, propagated by photons. Each photon carries quanta of energy. A harmonic oscillator, according to Planck, has energies only in a discrete set of equally-spaced levels, extension of an idea to electromagnetic field.

The reason to assume that the electromagnetic field is a quantum operator was simple. Scientists found that, at short distances, or in the presence of very strong fields, the classical field theory of electromagnetism breaks down. The quantum field reduces to the classical one under familiar circumstances, but differs sharply from it in some regimes of distance or energy.

In modern physics and quantum physics, we associate a field not just with waves but also with particles. An elementary particle is some type of coherent excitation of a quantum field. Thus, we associate an electromagnetic field with the fundamental particle, called the 'photon'. An electromagnetic field has discrete, evenly-spaced energy levels. One usually identifies these energy levels as different numbers of photons. The higher the energy level of a wave mode, the more photons there are. In this way, an electromagnetic wave acts as if it were made of photon carrier particles.

We will return to all the four fields several times in this book. Let us visit the other strong and weak fields to complete the present discussion. The weak field interaction occurs within the nucleus of an atom, and it is responsible for the change of neutrons and protons into each other in radioactive processes and stars. The energy released during radioactivity comes from the mass that's lost as the atom's nucleus decays.

Einstein determined that the amount of this energy (E) equals the mass lost (Δm) times the speed of light (c) squared, i.e., $E = \Delta mc^2$. In other words, energy and mass are two sides of the same coin. Furthermore, if one can create energy by a loss of mass, then one can also create mass from energy. The carrier particles of the weak interactions are the W+, W- and the Z bosons. The W's are electrically charged and the Z is neutral.

Weinberg and Salam predicted the existence of these spin-1 particles in addition to the photon in 1967 and unified the weak field with the electromagnetic field. Their theory also exhibits spontaneous symmetry breaking, which implies that, at high energies, all particles behave similarly but they appear to be different at low energies. In fact, it is known that the weak field and electromagnetic field become stronger at high energy, whereas the strong field described below becomes weaker and the three fields merge together around 10^{16} G eV of energy.

The strong field interactions also occur within the nucleus of an atom, which holds together the protons and neutrons, as well as the quarks within the protons and neutrons, inside the nucleus of an atom. The main difference between strong and electromagnetic interactions is that the carrier particles of the strong force, gluons, carry a color charge, but photons have no color charge. Two or more quarks in close proximity rapidly exchange gluons, and create a very strong 'color force field' binding the quarks together.

There are three color charges and three corresponding anti-color (complementary color) charges. Quarks constantly change their color charge as they exchange gluons with other quarks. Each quark has one of the three color charges. Each anti-quark has one of the three complementary color charges. Gluons carry color/anti-color pairs (which don't necessarily have to be the same color; i.e., red / anti-blue gluons are acceptable).

Although there are nine possible combinations of color/anti-color pairs, one of these combinations is absent due to symmetry considerations. A gluon can effectively carry one of eight possible color/anti-color combinations. At high energies, the strong field becomes weaker, making quarks and gluons behave almost like free particles.

The Standard Model combines electromagnetic interactions and weak interactions into a unified interaction called electroweak. The Standard Model includes a field for each type of the particles observed in laboratories, such as leptons and quarks. The quarks comprise protons and neutrons that make up the nuclei of ordinary atoms. The exchange of carrier particles produces the forces between these particles. Quantum electromagnetism describes the photon and its interactions with charged particles. Quantum Yang-Mills theory describes W and Z bosons and gluons (the carriers of weak and strong nuclear forces) and their interactions. The combination of all these theories makes up a single large theory called the 'Standard Model' of particle interactions, which is a quantum gauge theory.

To sum up, the Standard Model unifies all three non-gravitational fields, namely, the electromagnetic, strong, and weak fields. At a specific high energy, they do appear to have the same strength. However, attempts to include gravity have failed thus far.

- Future of the Standard Model

A few years ago, scientists called the proton an 'elementary' particle, and now they've discovered even smaller particles called 'quarks'. It is possible that scientists will find particles even smaller than the current 'fundamental' particles, when they are able to smash atoms using particles energy greater than 1000 GeV - the current energy limit for the particle accelerators. In fact, physicists recently discovered two new particles, designated as Ds mesons – Ds (2317) and Ds (2463) that contain a charm quark and an antimatter quark called antistrange. These particles have less than the expected mass, and are forcing physicists to reconsider how quarks interact.

By now, if you are feeling dizzy with the names of so many particles and the rules governing their behavior, you are not alone. Many scientists feel that, despite the many successes of the Standard Model, it does not tell the complete story. The Standard Model, which is based on 'quantum Yang-Mills theory', links the behavior of particles to structures found in geometry.

Despite its elegance and successes, we still do not understand the real physics behind it. The present Standard Model cannot predict various properties, such as the mass of a particle, and it fails to include the gravitational field. Recent findings that the neutrino has mass, suggest that the Standard Model needs a revision. Are we going to discover more particles and fields? What does the future hold for the Standard Model? These are interesting questions indeed!

- More Dimensions for Unification

The idea of the extra dimensions came from attempts to unify the different fields or forces of nature. Back in 1920, mathematicians found that five dimensions, instead of the four dimensions of space-time, helped them to reconcile electromagnetism with gravity. Specifically, in the 1920s, Kaluza and Klein (KK) attempted to show that one might account for the electromagnetic force by a fifth dimension. It was on the same lines as Einstein's assertion that gravity arose from the curvature of the four-dimensional fabric of space-time. They assumed that the fifth dimension must roll smaller than an atom. This was because one never sees such effects of the dimension at normal energies and distances.

Increasing dimensions is like climbing a hill to look down on a two-dimensional battlefield and seeing all the parts of the battle plan fitting together. In Kaluza-Klein (KK) theory, each point of normal space is actually a loop in this fifth dimension. A charged particle, even at rest in normal space, travels continuously around the loop like a hamster in a wheel. The electric charge is actually in motion in this hidden dimension. There are a number of satisfying connections between this motion and classical electromagnetism. For example, one can get the law of conservation of electric charge from the application of Newton's law of motion, i.e., for every action along the rolled-up dimension there is a reaction.

In recent years, scientists have postulated more dimensions. Extension of the KK theory postulated even more dimensions to include strong and weak forces. An electric charge in the electromagnetic force might be motion in the fifth, hidden, rolled-up dimension. In the same manner, quark properties, such as flavor in the weak force and color in the strong field, might be orbital dances in multidimensional KK loops. To take into account the different interactions observed in nature, one needs to provide the particles with more degrees of freedom than just their position and velocity. These degrees of freedom include mass, electric charge, color charge, and spin, etc.

In fact, string theory requires a string to vibrate in 10 dimensions (six extra dimensions too small to detect) to be mathematically consistent. Many scientists believe that string theory is a good candidate for the unified theory. String theorists are quite optimistic about it. String theory requires extra dimensions for consistency. It also seems to have mathematical elegance, consistency and a beautiful structure. However, scientists are not yet sure if it describes nature correctly.

I.   String Theory – A Candidate for Unified Theory

The basic idea in string theory is simple. We postulate that nature at the fundamental level has only one string-like object - not many types of elementary particles and fields. Just as a musical string produces distinct notes when it vibrates, this basic string can vibrate in different modes, and one can view each mode as an elementary particle. That is how string theory becomes a model of elementary particles. One cannot describe a string in terms of more fundamental constituents.

One might describe such a string classically by giving the location of an object extended like a (straight or curved) line in space at a given time. The string could be closed like a loop, or open with two ends, as illustrated in Fig. 1.11

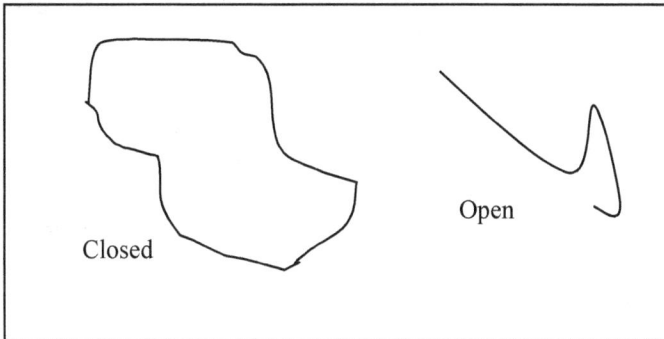

Fig. 1.11 Closed and Open Strings

String theorists describe a fundamental quantum string - an object with a finite spatial extent. New properties with profound implications for particle physics and cosmology appear when we apply the rules of quantum mechanics to a vibrating string, with vibrations propagating at the speed of light.

According to Veneziano, the father of string theory, the following properties of quantum strings are significant:

1) *Quantum strings have a finite length not smaller than the Planck length – a limit set by Heisenberg's Uncertainty Principle*
2) *Quantum strings may acquire up to two units of h of angular momentum without gaining any mass*
3) *Quantum strings require extra dimensions for equations describing vibrations to be consistent*
4) *Physical constants, such as the gravitational constant and Coulomb's constant, do not have fixed values, but occur as fields that can adjust their value dynamically. One such field is dilaton.*

The mathematical equations describing strings are consistent. Depending on whether a string is open or closed, it traces out a tube or a sheet as the string moves through time. String theorists believe that particles need additional degrees of freedom to take into account different interactions observed in nature.

The basis of the Superstring Theory is the concept of supersymmetry. It describes both types of particles, bosons, or carrier (integer spin) that carry force (e.g., photon, gluon, and even graviton) and fermions (odd half-integer spins) that constitute matter (e.g. electrons and quarks). String theorists proposed in 1984 the superstring idea The Universe as made up of one-dimensional strings vibrating in a background of nine dimensions of space and one of time. Today's superstring theories, which view the fundamental building blocks of matter as manifestations of tiny pieces of vibrating 'string', require 10 dimensions. The six extra dimensions roll up with a radius curvature of 10^{-35} m – the Planck size scale at which gravity becomes comparable in strength to the other forces of nature.

Six extra dimensions imply that we can have many more adjustable parameters. For example, we can wrap one extra dimension in a circle, and bundle the other extra dimensions in many different shapes or topologies, such as a sphere, one or two doughnut shapes joined together, etc. The solution to equations for different space geometry is not unique.

This has led to the development of five different string theories. However, string theorists believe that these theories are different parts of one theory that has not yet been discovered. Some string theorists call the potential energy associated with each solution, the 'vacuum energy' – the energy of space-time when the usual four dimensions do not have any matter or fields. The geometry of the smaller dimensions adjusts to minimize this energy.

In 1995, Witten and Townsend added one more space dimension to create M-theory in an attempt to unify the several string theories that physicists had dreamt up. The string theorists have also extended the notion of the one-dimensional string to multi-dimensional membranes called 'branes'. One-dimensional strings are one-branes, two-dimensional are two-branes, and objects with p-space dimensions are p-branes.

The latest belief is that the ends of the strings tie to a brane, except for strings representing gravitons. The string representing a graviton is free to move out of the brane in which our Universe exists. It leaks into the higher dimensions. String theorists attempt to explain the acceleration of space due to the weakening of the gravity field at very large distances due to such leakage of gravitons.

String theory has had its difficulties. For example, supergravity theory, based on supersymmetry related to spin, was popular for a while, since it could cancel infinities resulting from energy due to quantum fluctuations. However, it required superpartners for all the fundamental particles. Then, attention turned to superstrings, as they suspected that there might still be infinities lurking in supergravity. In superstrings, ripples correspond to fermions and bosons, and the ground-state energies or infinities resulting from the quantum fluctuation cancel out exactly.

The excitement is again building up after the discovery of several types of dualities that provide a direct link between several string theories. Intense research activity is starting again. String theorists made substantial progress in understanding the relationship and dualities between different string theories. For example, T-duality holds that small and extra large are equivalent. We are about to visit how the Universe originated from a Big Bang and the issues related to the pre-Big Bang era. String theory also proposes two scenarios for the pre-Big Bang era. One of these models uses T-duality with the symmetry of time reversal.

According to Randall and Sundrum, gravity's comparative weakness would be perfectly understandable, if particles called gravitons could leak-off a brane into a five-dimensional bulk. Alternatively, gravitons could be leaking across the bulk into our own brane (weak brane) from an extra-dimensional brane nearby (gravity brane). Despite recent progress, and despite its claim of predicting graviton and the promise of unifying gravity with quantum field, string theory has not yet quite succeeded in providing conclusive answers.

Some scientists put string theory in the realm of metaphysics instead of physics, until some experimentally verifiable propositions supporting the theory can be found. To save extra-dimensional physics from fiction, we need real-world evidence to support the brane world concept. An excellent reference on string theory is Brian Greene's book, The Elegant Universe.

Scientists are pinning their hopes on the underground particle accelerator, called the Large Hadron Collider (LHC), at Europe's CERN laboratory on the French-Swiss border. It is the world's largest machine and is housed in a 27km-long circular tunnel. The $10-billion giant underground laboratory would smash particles together at super-fast speeds in a bid to unlock the secrets of the Universe. The accelerator would smash protons together in a 27km-long ringed tunnel, with enough energy to spawn subatomic particles that have momentum in the extra dimensions. We would see this momentum as extra mass. If the LHC produces classes of new particles with the same charges as normal particles, but heavier by certain amounts, it could indicate a fifth dimension.

In this LHC machine, over 1,000 powerful 'dipole' magnets occupy the subterranean tunnel, which houses the accelerator. These magnets carry two beams of particles around the ring at speeds close to the speed of light. The eight magnet assemblies, called 'inner triplets', consist of three 'quadrupole' magnets, which are cooled using superfluid helium at $1.9°$ Kelvin ($-271°C$) inside a vacuum vessel.

The Compact Muon Solenoid (CMS) experiment, along with the rival experiment, Atlas, will seek to identify the elusive Higgs boson - important to the Standard Model of particle interactions - look for so-called supersymmetric particles, and seek out the existence of extra dimensions.

Scientists hope that it will shed light on the fundamental questions in physics, and probe the limits of physics to understand the true nature of matter and the origin of the Universe. The lab's 20 European member countries, as well as observer states like the United States and Japan, contribute to CERN's annual budget of about $940 million. In 2009, engineers have made two stable proton beams circulate in opposite directions around the machine. This progress is encouraging.

In March 2010, the machine created two beams of protons, each with the energy of 3.5 trillion electron volts. This effort breaks the prior record, set by the LHC in December 2009, of just over a trillion electron volts in each beam. The LHC will now aim to smash those two beams together, hoping to create new particles that will give us insight into the most fundamental workings of physics.

1.9 Summary & Concluding Remarks

This first phase of the journey revealed the mysteries of science. During the journey, we've met several scientists. Let us summarize what we discovered during this phase of our journey.

Newton explained motion and gravity; Maxwell explained electromagnetism and light waves; Einstein gave the world his theories of relativity, changing the notions of time, space, and gravity; Planck defined the quanta of energy; and Schrödinger introduced quantum mechanics. Thus came, in quick succession, remarkable scientific discoveries over a short period of human history.

While crossing various milestones, we've learned more about how these scientists discovered various physical laws. Scientists have become bold and they now want to develop a single theory to explain the origin of the Universe, the formation of stars, galaxies and black holes. In this long first chapter, we highlight the major achievements in science.

Newton's laws of motion and gravity were perhaps the first significant developments in the history of science. These laws were very appealing because of their universality. For example, the same law of gravity that governs the fall of an apple also applies to the orbits of the planets. Newton's laws of motion could explain a vast range of phenomena related to motion. Many thought that physics would end there. However, scientists soon found these laws inadequate, and were uncomfortable with the notion of action at a distance in which gravity acted instantaneously across the Universe.

Next, Maxwell formulated the theory of electromagnetism, based on physical laws for the electromagnetic phenomenon, and put it in concise mathematical form, now known as Maxwell's equations. These equations gave a complete description of electromagnetic phenomena, and predicted the propagation of electromagnetic waves, ushering in the communication revolution.

Soon, Einstein's two theories extended Newton's laws for motion of objects close to the speed of light. Einstein's special and general theories of relativity accounted for the discrepancies in planetary motion and replaced action at a distance with the curvature of space-time. General relativity provided deep insight into the origin and evolution of the Universe, e.g., the Big Bang, the evolution of stars and galaxies, existence of black holes, and the gravitational lenses. Einstein's theories are adequate, so long as we do not apply them to the subatomic or micro-scale.

Scientists developed quantum mechanics and quantum field theory to explain the atomic and the subatomic phenomena (micro). On the scale at which we live our normal lives, the Universe appears continuous. According to quantum mechanics, however, the Universe, when observed on a micro-scale, is discontinuous and divided into discrete bits. Thus, quantum mechanics treats matter and energy as composed of particles and waves with a certain quanta of energy.

The Standard Model and the unification of three non-gravitational fields were directly the results of developments in quantum mechanics. Quantum mechanics proved extremely successful in explaining microscopic subatomic phenomena; where the effect of gravity is weak, one can ignore it. Quantum field theory, which includes the theory of special relativity, has worked extremely well to describe the properties and observed behaviors of elementary particles. It is accurate to about one part in about 10^{11}, and general relativity has now been tested to be correct to one part in 10^{14}, limited by the accuracy of clocks.

Both relativity and quantum theories were necessary. Relativity theories ignore quantum effects, predominant at micro-scale, and quantum theory cannot account for the gravitational field. Both theories in their domains of application, however, have proved to be very accurate. Either theory loses its predictive power whenever it becomes impossible to ignore the other. For example, relativity and quantum mechanics fail to converge for the description of the black hole, which has immense gravity but extremely small size.

The fact that we cannot reconcile both these theories seems to indicate that something is not quite right, and new discoveries are yet to come. We must resolve the conflict between these two beautiful theories of the twentieth century. To understand the events leading up to the Big Bang and black holes, we must unify quantum mechanics and Einstein's gravitational theory by looking at space at the quantum level. Scientists are attempting to develop a theory called 'quantum gravity' to build a bridge between quantum mechanics and Einstein's general relativity.

The other strong contender, string theory, talks about quantum strings as the fundamental component of the Universe, and not the point-like character of the most fundamental objects in this Universe. With this premise, one may be able to obtain a unified theory that reconciles the current disparities in existing theories. . String theorists believe that Einstein's theory will reemerge as a low-energy limit from the superstring theory. There is lot of excitement, intense research activity, and high expectations from string theory, but the jury is still out on whether it can fulfill such high expectations.

Chapter 2
Unfolding the Mysteries of the Universe

2.1 Introduction

We've revealed the mysteries of science in the previous chapter. By now, we should have a good understanding of what science is all about. It is now time to see how science can unfold the mysteries of the Universe. We unfold a truly fascinating story. We visit the most important contributions of science, and understand our Universe – its size, age, origin, and evolution.

2.2 Universe - Age, Size and Contents

The Universe is indeed quite old, and its present size is vast. Let us start with the age and the size of the Universe, as determined by science. From old star clusters and recent data from the WMAP space probe, scientists estimate that our Universe is 13.7 billion years old $(4.32 \times 10^{+17} \text{sec})$ with a 1% margin of error. In other words, it takes light 27.4 billion years to go from one end to the other end $(\sim 1.89 \times 10^{+21} \text{m})$. The speed of light is 186,000 miles per sec $(\sim 3 \times 10^{+8} \text{m/s})$ and one year has 31,556,926 seconds.

As regards the size of the Universe, one might think that the corresponding width of the observable Universe would be 27.4 billion light years $(\sim 2.6 \times 10^{+26} \text{m})$. However, during this 13.7 billion year period, the Universe has also been expanding. The distance between galaxies has grown by a factor of about 1000. The radius of the observable Universe has increased by a factor of about 100,000, because light outpaces the expansion. All the pieces add up to 78 billion light years. According to Cornish, the diameter of the Universe is about 156 billion light years, based on a view going 90 percent of the way back in time, so it might be slightly larger.

Most astronomers agree that the value of the Hubble constant, a number that measures the expansion rate and age of the Universe, is about 71 kilometers per second per Mega parsec (one Mega parsec is 3.2 million light years). A recent project to create an easier way to measure cosmic distances suggests that the Hubble constant is actually 15 percent smaller than what other studies have found.

Astronomers have observed over 10 billion galaxies, but they think that the Universe has over 125 billion galaxies of different shapes and sizes. Recently, they detected a galaxy near the edge of the Universe, over 13 billion light years away. The galaxy, named IOK-1, is so far away that the light waves that reach Earth depict it as the system of stars that existed shortly after the Big Bang .

In May 2007, astronomers also released, at the American Astronomical Society Meeting in Hawaii, an awesome picture of a spiral galaxy M81, 11.6 million light years away, taken by the Hubble Telescope. This beautiful galaxy, tilted at an oblique angle to our line of sight, gives a 'bird's eye view' of the spiral structure. The galaxy is similar to our Milky Way. "The view we have of M81 is similar to what an astronomer in Andromeda would see if they looked at the Milky Way," explained Zezas, an astronomer at the Smithsonian Center for Astrophysics. It is interesting to note that the galaxy has a black hole located at its center, which contains 1/2 % of the mass of the galaxy.

Our galaxy, the Milky Way, is a spiral galaxy, a collection of about 100 billion stars - Sun being one of them. The Milky Way also has a black hole in the center. Our galaxy has curving arms spreading out from the core, and our solar system is on one of the spiral arms. Our galaxy is about 200,000 light years wide.

The astronomers discovered new stars in 2007 in this galaxy. Images from the new European telescope in La Silla, Chile, reveal a previously unknown rich cluster of stars in the inner parts of the Milky Way. This closely packed star family consists of about 100,000 stars, and it is located some 30,000 light years away. The Milky Way's star HE 1523 is the oldest-known star at 13.2 billion years old (the Big Bang occurred 13.7 billion years ago). They estimated the star's age from what is left of its radioactive elements, compared to stable 'anchor' elements, like europium, osmium, and iridium.

In 2009, Europe's billion-euro Herschel Space Observatory obtained a remarkable view of our galaxy. The images reveal, in exquisite detail, the dense, contorted clouds of cold gas that are collapsing to form new stars. Herschel, which has the largest mirror ever put on an orbiting telescope, was launched in May as a flagship mission of the European Space Agency.

Our nearest neighboring galaxy, Andromeda, is two million light years away from our galaxy. When we look at it now, through a telescope, we are observing events that took place two million years ago – when humans had not yet appeared on Earth. A computer simulation study suggests that our galaxy might merge with Andromeda in about 7 billion years, forming one large elliptical galaxy.

Most galaxies have billions of stars. Our Sun is one such star in our galaxy, which is about 4.5 billion years old. A star has nuclear reactions on its surface, but a planet does not. The next star nearest to the Sun is 25 million miles away. Many of these stars have planets orbiting them. The new computer models suggest that our solar system is hurtling through space angled nearly perpendicular to the plane of the Milky Way. According to the Merav Opher at George Mason University in Virginia, our solar system is sailing through the galaxy.

Our planet Earth in the solar system, orbiting the Sun, is 93 million miles away from the Sun. Earth is about 2/3 distance away from the center of our galaxy. The light from the Sun takes 8 minutes to reach us. A new calculation in 2008 predicted that, in 7.6 billion years, the Sun would swallow up the Earth. This answers the question of whether the Sun's gravitational pull will have weakened enough for Earth to escape final destruction or not.

In July 2005, astronomers discovered the tenth planet in our solar system - called 2003 UB313, with a diameter that's 200 miles larger than that of Pluto's. It also has a companion Moon. It is currently 14.5 billion km (9 billion miles) away, about 97 times further away than the Earth is from the Sun. It has a highly elliptical orbit, tilted at roughly 45 degrees above the orbital plane of the other planets.

An expert panel recently proposed that astronomers should stop using the name 'planet' for every object discovered in the solar system. In August 2006, a meeting of the International Astronomical Union in

Prague first decided on 12 planets in our solar system, shown in Fig. 2.1. However, they finally decided on only eight planets in the solar system. The decision establishes three main categories of objects in our solar system:

- *Planets:* Mercury, Venus, Earth, Mars, Jupiter, Saturn, Uranus, and Neptune.
- *Dwarf Planets:* Pluto and any other round object that 'has not cleared.the neighborhood around its orbit, and is not a satellite.'
- *Small solar system bodies:* All other objects orbiting the Sun are smaller solar system bodies.

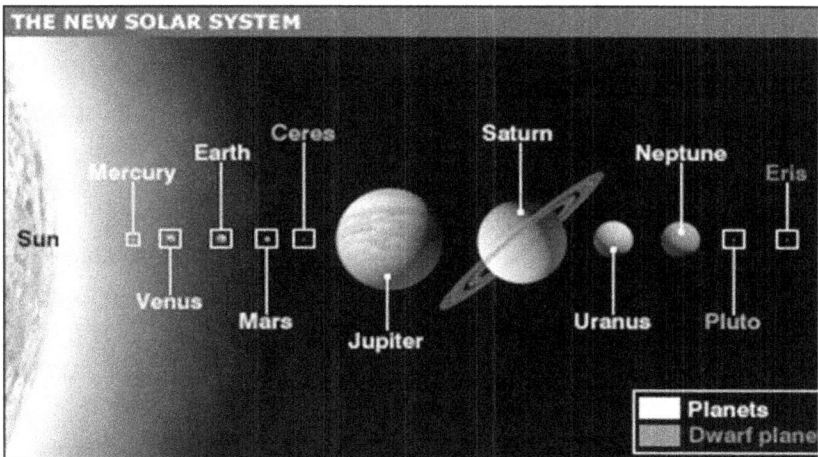

Fig. 2.1 - Twelve proposed planets - Ceres, Pluto, Charon, and 2003 UB313 are barely visible

Scientists have perfected techniques to identify planets. A space telescope aboard a spacecraft, Corot, launched from Kazakstan, is monitoring about 120,000 stars for tiny dips in brightness that result from planets passing across their faces. On average, astronomers are discovering a new planet every two weeks. When a planet orbits its star, it also exerts a gravitational pull. This causes the parent star to 'wobble' around its centre of mass. The High Accuracy Radial velocity Planet

Searcher (HARPS) spectrograph is able to measure this wobble to a very high precision

Using these techniques, astronomers have been discovering new planets in many star systems. Since the discovery in 1995 of a planet around star 51 Pegasi by Michel Mayor and his colleague Didier Queloz, astronomers have found more than 270 planets, mostly around Sun-like stars. The majority of these planets are gas giants, a bit like Jupiter or Saturn in our own Solar System. Current data shows that about one in 14 stars has this kind of planet. In 2008, NASA announced that scientists have detected organic molecules, methane, and water in the atmosphere of an extra solar planet known as HD 189733b. The planet is located 63 light years away in the constellation Vulpecula. It is so massive and so hot that it would be an unlikely host for life.

In 2008, astronomers also identified a trio of so-called 'super-Earths' - rocky planets between two and 10 times the mass of Earth. They detected these three new planets using the HARPS instrument at the La Silla Observatory in central Chile. The star that the planets circle is slightly smaller than our Sun, and is located 42 light years away near the southern Doradus and Pictor constellations.

We are discovering more planets at a rapid rate. Recent data from NASA Kepler's space telescope identified over 1200 planets outside our solar system around 2,000 light years away, over a period of 4 months, looking at 1/400 of the night sky. For the first time, the scientists using this space telescope, also discovered six planets made of a mix of rock and gases orbiting a single sun-like star, known as Kepler-11, located approximately 2,000 light years from Earth.

The Kepler telescope, shown in Fig.2.2, searches for exoplanets, looking into a small, fixed patch of the sky in the direction of the constellations Cygnus and Lyra. The basic idea for identifying the planets is the same as pointed out earlier. The Kepler telescope looks for the minuscule dimming of light that occurs, when an exoplanet passes in front of its host star. The "candidate" planets spotted by Kepler are confirmed by ground-based observations to confirm their existence.

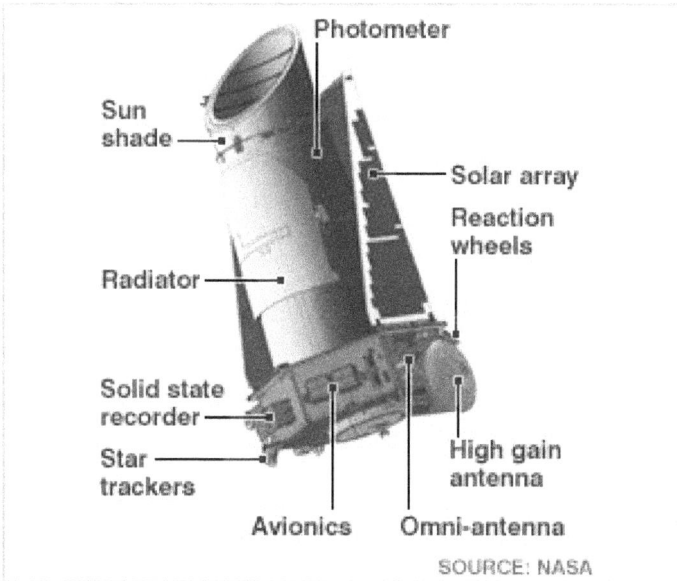

Fig. 2.2. Kepler Telescope (NASA)

Out of these 1,200 planets, 54 were in the habitable region with the possibility of the presence of water , and 6 of these are about the size of earth. But Kepler can only identify the presence of a planet. Future spacrafts like Darwin will look more closely at some of these planets, and search for existence of intelligent life. However, this task is quite formidable. In any case, even if we find intelligent life, because of the distances involved it would not be easy to communicate as signal traveling at the speed of light light would take 2,000 years to reach there. Fortunately, this vast distance also puts to rest any ideas about alien attack on earth.

Finally, astronomers are asking the question if every star has planets and, if yes, how many? We do not yet know the answer, but we are making progress.

2.3 Scale Problem for Space, Time & Energy

The scale on which we live our lives, the distance we cover in a step, and the time span of our life, dictate what we consider normal. To an ant (living on a different size and time scale) crawling over our body, the body would look extremely vast, and its shape abnormal. If the ant could talk, it would describe its world as vast as the Universe appears to us. Various objects in this Universe exist over a vast scale, when we observe on a scale dictated by our senses. (Fig. 2.3).

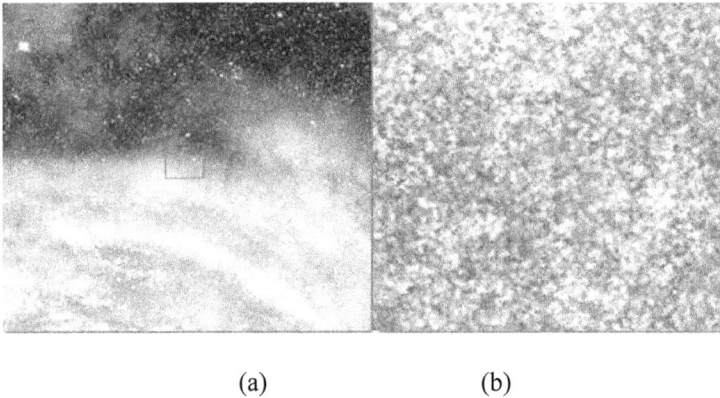

(a)   (b)

Fig. 2.3 (a) Our Galaxy (inside marked box) at 100,000 light yrs distance; (b) Quarks at 10^{-16}m

Light takes a fraction of a second to travel across the entire United States, eight minutes to travel from the Sun to the Earth, but it takes more than 10^{17} seconds (13.7 billion years) to travel from the farthest corner of the Universe. The Universe would shrink and appear to be of reasonable size to anyone, or any particle, that could travel at a speed that's millions of times the speed of light.

Similarly, considering the size of the nucleus (10^{-14}m), atom (10^{-10}m), DNA (10^{-8}m), cell (10^{-4}m), animal (10m), stars, galaxies, and the edge of the observable Universe, we cover distance scales all the way up to 10^{26}m. Compared to the size of atoms, we are huge, but compared to the size of the Universe, we seem insignificant. A planet like our Earth appears infinitesimal compared to the size of the Universe. Even our entire solar system is smaller than a speck of dust in this Universe.

Objects and distances in this Universe on one end of the spectrum are so large, but on the other end, they are so small. They can start from 10 million light years (10^{23}m.) at space, and end at 100 atom/meter (10^{-16}m.) on Earth

NASA recently described, through beautiful pictures, a journey that starts and ends in distances which are difficult for the mind to capture.

- At 100.000 light years (10^{2}1m), we can barely see our galaxy. (Fig. 2.3)
- At 10.000 light years (10^{20}m), we can see the stars of our galaxy.
- At 1 light year (10^{16}m), we can barely see the Sun.
- At 100 billion Km (10^{14}m), we can see the solar system.
- At 100,000 Km (10^{8}m), our Earth still looks small.
- At 1.000 Km (10^{6}m), we can recognize coastal areas in USA.
- At 1 Km (10^{4}m), we can see distinct places.
- At 10 cm (10^{-1}m), we can see the leaves.
- At 100 micron (10^{-4}m), we can see the cells.
- At 1 micron (10^{-6}m), we can see the cell itself.
- At 100 angstrom (10^{-8}m), we can see the DNA chain.
- At 1 angstrom (10^{-10}m), we can see the carbon atom.
- At 10 Pico meter (10^{-11}m), we can see the electron in the atom.
- At 100 Fermi (10^{-13}m), we can see the inside of an atom.
- At one Fermi (10^{-15}m), we can see the surface of a neutron.
- At 100-atom-m (10^{-16}m), we can see the quark. (Fig. 2.3)

In fact, the distance, time and energy scales at the lowest and the highest end of the spectrum are difficult to visualize. As stated, this is due to the normal scales (dictated by the limitations of our physical senses) that we use in our daily life. In visualizing different scales, the power of 10 can also be misleading. For example, if we said that the observable Universe is $\sim 4.32 \times 10^{17}$ seconds old, it might not register that we are talking about time that's close to 13.7 billion years.

On the lowest end of time, we note that Planck time is $\sim 10^{-43}$ sec. It is hard to visualize this interval. We know what a second feels like, but we cannot relate 10^{-43} seconds to any physical experience. On the low end of the distance scale, we have the Planck length of $\sim 10^{-35}$ meters, and we have difficulty visualizing such small distances. It is 10^{-25}th times the size of an atom ($\sim 10^{-10}$ m), and one could place almost a million atoms on a needlepoint.

We have developed detectors that allow us to explore the Universe on different scales. These detectors include particle accelerators, microscopes, and telescopes, like the Hubble telescope in space. The methods to observe differ, depending on the size and distance of the object we want to view.

Regarding matter and energy in this Universe, there are over 120 billion galaxies in the Universe. Each galaxy has, on the average, billions of stars, and many of these stars must have planets. Regarding energy, the explosions in stars and the fireworks in galaxies dwarf the energy produced by the Sun.

The energy produced from nuclear reaction in the Sun is far more than the energy produced by millions of hydrogen bombs. Yet, all ordinary matter and energy constitutes only 4.4% of the Universe. Surprisingly enough, we have no clues as to the exact nature of the remaining dark matter (23%) and dark energy (73%). The ultimate goal of science is to develop the final theory that would explain every phenomenon in the Universe, irrespective of space, time, or energy scale.

2.4 Origin of the Universe

How and when did our Universe originate? Astronomers and cosmologists refer to the events that took place billions of years ago. We lose interest and ask why we should care about such a distant past. A few simple examples should make us pay attention to this distant past. It is indeed amazing that the hydrogen atoms in the water (H_2O) that we drink everyday were formed more than 13 billion years ago, just three minutes after the Big Bang. The lithium atoms in anti-depressants were also provided to humanity after the first three minutes, perhaps anticipating our depression while trying to understand the Universe.

- Big Bang Model

Scientists and Cosmologists claim that our Universe originated with a Big Bang. Hoyle had originally given the name 'Big Bang' to ridicule this idea. Now, the Big Bang model is indeed a widely accepted theory about the origin of the Universe. Recent observations from various space probes seem to lend support to this theory. The Big Bang theory describes an expansion process, which started with an explosion 13.7 billion years ago. It mushroomed from a hot, infinitely small state - almost a point singularity - created new space, and evolved into the current Universe. Since the process was like an explosion, cosmologists called it the Big Bang. The Big Bang was unbelievably energetic, but no one knows for sure where the energy came from. Part of that energy was subsequently transformed into matter while the Universe evolved, as different atoms - the primary constituents of the matter - were created.

Cosmology is the branch of science that is concerned with the study of the Universe: its origin, its evolution, its expansion rate, its mass, and its dark matter and dark energy, etc. A cosmologist studies how the Universe was created and how its structure was determined. It became a mature science only about thirty years ago. Cosmology at this stage, nevertheless, is far from a well-developed science. On the theoretical side, they are trying to further develop several theories, such as the inflation theory, to explain certain observed characteristics of the Universe. On the observational side, cosmologists are analyzing a lot of gathered data to learn about the Universe.

Some cosmologists believe that the laws of physics might explain the origin of the Universe, but they rarely discuss the origin of these laws. Cosmologists base their belief in the laws of physics on the success of these laws in explaining the astronomers' observations concerning various aspects of the Universe. Astronomers have also studied the spectra of light emitted by distant stars billions of years ago, and have found no evidence so far that the laws of physics have changed over this period

How credible is the Big Bang model? Understanding of the Big Bang comes from particle physics, from the mathematical model of an expanding Universe based on general relativity, and from the observations of various space probes. For example, physicists working in particle physics think that a ten billionth of a second after the Big Bang, the size of our Universe was that of a typical living room. It had an unbelievably high temperature, around billions and billions of degrees ($\sim 10^{20}$K). Matter and energy were exchangeable; particles and antiparticles were created and annihilated. Somehow, the symmetry was then broken, and for every billion antiparticles, there were a billion plus one particles.

Physicists go on to say that one microsecond after the Big Bang, the size of the newborn Universe was only a few kilometers, and temperature had dropped to several billion degrees. In this special state, the surplus quarks and electrons - the basic building blocks of matter - floated freely in an incredibly hot, dense soup. These quarks bonded together into protons and neutrons, as the Universe expanded and cooled. Eventually, these protons and neutrons bound with electrons, forming the matter that we see today.

In fact, scientists are trying to create this primordial soup in particle accelerators, called quark-gluon plasma. Scientists do find anti-electrons (positrons) in the cosmic rays coming from the sky, but their search for the anti-nuclei in space has not been successful. The AMS-2 experiment on the International Space Station is designed to search for any leftover traces of Big Bang antimatter

Einstein's equations give us a mathematical model for the expansion rate of the Universe at different stages and times, given the energy density of matter and radiation at that time. The matter and the radiation density of the early Universe are estimated from the ancient light that reaches us from the past in our night skies. More discoveries, both observational and theoretical, continue about the origin and evolution of the Universe.

The Hubble Space Telescope has provided wonderful pictures all the way to the edge of the Universe. A recent Hubble Space Telescope study, designed to study evolution in the stellar Universe, took a mosaic of photos combining 78 separate exposures. This mosaic includes more than 40,000 galaxies in a patch of sky, about the size of the full Moon. Observations also confirm the existence of black holes in each galaxy.

On the 18th anniversary of the Hubble Space Telescope's launch, scientists released a series of 59 new photographs in 2008. It is the largest collection of Hubble images ever released together. These Hubble space images show galactic collisions in action and the variety of peculiar forms that the merging galaxies can take. The current view is that galaxy mergers are more common than previously believed. In fact, they were even more common in the early Universe than they are today, since the early Universe was smaller, so galaxies were closer together, and were thus more prone to collisions and mergers. The galaxies that appear isolated also show signs of past mergers in their internal structure.

- Support for the Big Bang Model

The observations of the Universe seem to support the Big Bang model for the origin of the Universe and the inflation theories. These observations have confirmed the abundance of hydrogen and helium, Cosmic Microwave Background (CMB) radiation, and the expansion of the Universe - predicted by the Big Bang. Astronomers believe that the CMB contains a great deal of information about the origin of the Universe and its structure that we see around us today. They also believe the CMB holds clues to the Universe's eventual fate, i.e., whether it is likely to go on expanding forever. Theory states that, 380,000 years after Big Bang, matter and radiation decoupled. Matter went on to form stars and galaxies; radiation spread out and cooled. The CMB radiation now shines at weak radio (microwave) wavelengths.

Scientists Mather and Smoot, who worked on NASA's COsmic Background Explorer (COBE) satellite that was launched in 1989, received the 2006 Nobel Prize in Physics. Their teams made the first precise measurements of the CMB, as they measured the temperature of this background radiation - a frigid 2.725 degrees above absolute zero. COBE showed that the CMB profile follows a predicted distribution - a so-called blackbody curve. COBE mapped tiny temperature fluctuations in CMB. These fluctuations (anisotropy) correspond to the early distribution of matter. The data informs scientists about the age, geometry, and fate of the cosmos.

In the second week of February 2005, NASA released observations from space probes, using NASA's Wilkinson Microwave Anisotropy Probe (WMAP), during a sweeping 12-month observation of the entire sky. This picture in Fig. 2.4 contains stunning details, a most important scientific achievement.

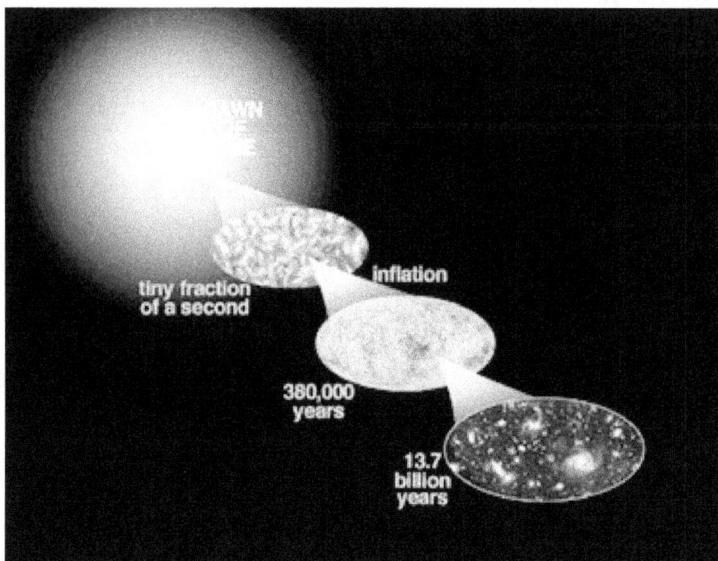

Fig. 2.4 – NASA-WMAP: Evolution of the Universe

The probe provided the best 'baby picture' of the Universe ever taken.The image captured by the WMAP is from the time when the temperature of the Universe became low enough for atoms to form, allowing light to travel great distances to reach us. It is similar to the surface of the clouds that we see on an overcast day. Light travels through the clouds, but we only see the detail on the cloud's surface. The light reaching us has been stretched out, since the Universe has stretched while expanding. Thus, the light that was once beyond gamma rays is now reaching us in the form of microwaves. Microwaves are part of the electromagnetic spectrum, just like light, but stretched out to a longer wavelength.

In this picture, scientists have also captured the afterglow of the Big Bang, called the cosmic microwave background. They show that the temperature fluctuations were shaped by acoustic waves for several thousand years. The latest observations clearly show that the patterns in the Big Bang afterglow were frozen in place only 380,000 years after the Big Bang. These patterns are tiny temperature differences within this extraordinarily evenly dispersed microwave light bathing the Universe, which now averages a frigid 2.73 degrees above absolute zero.

WMAP can resolve slight temperature fluctuations, which vary by only millionths of a degree. Theories about the evolution of the Universe predict these temperature patterns. The WMAP team compared these predicted fingerprints with the observed unique 'fingerprint' of patterns imprinted on this ancient light and found a match. Observations of the cosmic microwave background are expected to reveal new insights into the theory of inflation and the nature of dark energy.

A new unified understanding of the Universe, reported by the WMAP team, can be summarized as follows:

1) The visible Universe is 13.7 billion years old with a 1%.margin of error. Note that there could be objects that are more distant. Such objects will forever remain unknown to us. Interaction with such remote parts of the Universe would require some exotic means of travel, such as to travel faster than light. However, according to Einstein's special theory of relativity, accelerating an object to a velocity beyond the speed of light is impossible, as it would require an infinite amount of energy.

2) Light in the WMAP picture appears 380,000 years after the Big Bang.

3) The first stars ignited 200 million years after the Big Bang. Matter in the Universe condensed by gravity, until the first stars ignited. The WMAP has detected this event at about 200 million years after the Big Bang. The WMAP does not see the light of the first stars directly, but has detected a polarized signal that is the signature of the energy released by the first stars. WMAP data analysis starts with the temperature fluctuations in the oldest light in the Universe, as captured by the WMAP.

The temperature fluctuations correspond to the slight clumping of material in the infant Universe, which ultimately led to the vast structures of galaxies we see today. Then, matter condenses as gravity pulls matter from regions of lower density to regions of higher density. It then captures the era of the first stars, 200 million years after the Big Bang. Gas has condensed and heated up to temperatures high enough to initiate nuclear fusion, the engine of stars. More stars keep turning on, and galaxies are formed. Finally, it leads to the modern era, with billions upon billions of stars and galaxies.

4) Content of the Universe: The WMAP indicates that the Universe constitutes of 4% atoms, 23% cold dark matter, 73% dark energy, as illustrated in Fig. 2.5.

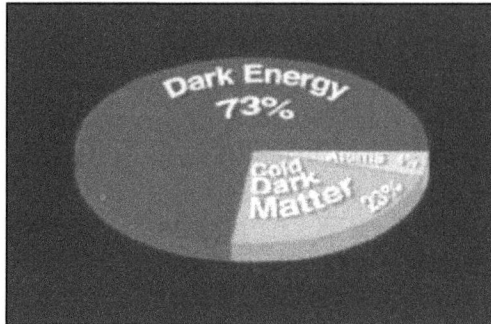

Fig. 2.5 – Contents of the Universe (NASA- WMAP)

The WMAP determined that about 4.4% of the mass and energy of the Universe is contained in atoms (protons and neutrons). All of life is made from a portion of this 4.4%.

5) Expansion rate (Hubble constant) value: Ho = 71 km/sec/Mpc (with a margin of error of about 5%)

6) New evidence for the Inflation theory (in polarized signal) According to the theory that fits with WMAP data, the Universe will expand forever. If it changes with time, or if other unknown and unexpected things happen in the Universe, this conclusion could change.

- Formation of Matter, Stars and Galaxies

After looking at the observations that support the Big Bang model and other evidence, let us construct the sequence of the events after the Big Bang. We can paint the following simple picture of the expansion, formation of matter, stars, and galaxies.

A ten billionth of a second after the Big Bang, the size of our Universe is that of a typical living room, with an unbelievably high temperature around billions and billions of degrees ($\sim 10^{20}$K). Matter and energy are exchangeable; particles and antiparticles are created and annihilated.

Somehow, the symmetry is broken. For every billion antiparticles, there are a billion plus one particles. As the Universe expands, its temperature keeps falling; particles and antiparticles keep annihilating each other, leaving a small surplus of matter particles.

Within one second after the Big Bang, the temperature falls to about a hundred billion degrees (a temperature that's thousands of times higher than the temperature currently existing in the center of the Sun). At such a high temperature, collisions between highly energetic gamma rays keep producing particle/antiparticle pairs (matter-antimatter), such as negative electrons and positive anti-electrons or positrons, or positive protons and negative anti-protons. Most of these particles again collide with their antiparticles, giving back the radiation. However, within the first second after the Big Bang, the protons, surplus neutrons, and electrons left behind are the ones that would turn into most of the matter we see today.

About three minutes after the Big Bang, the temperature drops to a little less than a billion degrees (still unbelievably hot), and protons and neutrons start to combine and electrons are captured by the nuclei. These reactions form hydrogen (H – one proton and one electron), deuterium (an isotope of hydrogen, with one proton and one neutron in its nucleus) and helium (He - with two protons and two neutrons) and a few lithium atoms. The Big Bang model thus explains the abundance of H, about 90%, and He, about 9% in the Universe. After three minutes, as the Universe keeps expanding, the temperature of the Universe falls rapidly, not leaving enough time for the helium to combine into heavier elements, since that requires very high temperatures.

For the next two hundred million years, the remaining 1% of ingredients - the rest of the atoms, besides H and He that are required for sustaining the present form of life - did not exist. Oxygen, carbon, iron, and silicon - none of these elements existed. The only existing atoms were hydrogen, helium, and lithium. By now, gravitational instabilities pull clouds of hydrogen and helium gases together to form stars and galaxies. Another recent theory paints a somewhat different picture, as it suggests the formation of black holes first from the swirling clouds of gas, which then create the stars and the galaxies.

As gravity pulls the gas together, hydrogen first converts into helium in the hot core of the stars. This helium has slightly less mass than the hydrogen, and the extra mass is converted to energy. This fusion process is also the basis of the hydrogen bomb. In fact, the nuclear reactions inside stars are the only possible source of energy, which has kept the stars hot for billions of years.

For example, through the fusion of hydrogen into helium in the Sun, about 4 million tons of matter is converted into energy every second in its core, where the temperature is about 15 million degrees Celsius. The Sun does not have enough mass to reach the higher temperature required for further fusion beyond hydrogen; it will die, fading away once the hydrogen supply is exhausted. Our Sun has been shining for 4.5 billion years and has enough hydrogen to keep going for another 5 billion years or so.

For stars more massive than the Sun, the fusion process continues. The interior of the massive stars provides the proper temperature environment for the fusion of the nuclei of light atoms and the subsequent formation of heavier atoms. As these stars run out of hydrogen fuel, their helium cores begin to contract and heat up under the influence of gravity, which overwhelms the repulsive force in the nuclei, raising the temperature even further.

The temperature gets high enough (~100 million degrees), so that the helium starts to fuse into beryllium, carbon, etc., again releasing energy in the process. Depending on the mass of the star, the fusion process goes on to create oxygen, neon, sulfur, silicon, and up to iron. Iron is the last stop, because instead of energy release, energy is required to fuse iron.

These stars and the garden-variety stars like the Sun live rather boring lives, churning out heat and light for billions of years. They eventually get old and die. However, before dying, these ordinary stars swell into so-called Red Giants. They begin casting off their external gas layers in violent eruptions. These beautiful cast-off layers are called planetary nebulae.

They were named planetary nebulae in the eighteenth century, because these gas clouds looked like the planet Saturn when seen through small telescopes by the astronomers. The Hubble Space Telescope has revealed their true nature. The name 'planetary nebula' is a misnomer. According to the intriguing, but beautiful, picture provided by the Hubble

Telescope, their shapes, sizes and colors, and the events taking place at the end of a star's life, are indeed very complex and hard to imagine.

Since extra energy is not available, eventually the core of the star collapses, triggering an explosion called a supernova. The resulting chaos during a supernova creates still more elements, all the way from iron to almost the last 120th element, which spew out into space by the force of the explosion. Eventually, the exploding debris merges with other clouds of gas to form new stars. These new stars have the heavier elements. The Sun is one of those stars, and its planets, such as our Earth, and the life on Earth, owe their existence to those earlier stars. The recycling process goes on.

For readers interested in scientific details, we describe the various stages of the expansion of the Universe, as follows:

1) Planck & Inflation Era   ($\sim 10^{-43}$sec $\sim 10^{32}$K)

Only one field exists. Soon after the Planck era, cosmologists believe that there was a period called 'inflation'. Planck and the inflationary era, filled with questions, are lumped together. The theoretical physics concerning these two periods is still in a state of flux.

2) Radiation fills the Universe   ($\sim 10^{-12}$sec $\sim 10^{32}$K)

At the end of the inflationary era, the Big Bang officially begins from a small, hot, dense quantum state. Gravity separates at 10^{-43} sec. At 10^{-25} sec and a temperature of 10^{27} K, quarks and anti-quarks form as the strong force separates from electroweak forces. The energy of the quantum field changes into radiation energy, a soup of photons, gluons, etc. The radiation energy density is very much larger than the matter part of the energy density.

3) Quarks outnumber anti-quarks   ($\sim 10^{-11}$sec)

Radiation creates pairs of particles and antiparticles, and these pairs annihilate back to radiation. As the Universe expands and cools, quarks and anti-quarks freeze out. Inexplicably, the Universe is left with more quarks than anti-quarks. All four forces have become distinct.

4) Weak nuclear bosons become massive   ($\sim 10^{-10}$sec)

Electromagnetic force separates from the weak force at 10^{-7} sec and at 10^{14} K. Average particle energy density drops below energy scale of weak nuclear force, where spontaneous symmetry break occurs. Weak nuclear bosons gain mass, slowing them down and restricting their force to a small range.

5) Quarks and gluons are confined $(\sim10^{-4}sec)$

Quarks and gluons, popping around at high speed, are confined at lower temperatures due to phase transformation into mesons and baryons (neutrons and protons). This would go on to produce the Universe that we observe today.

6) Proton-to-neutron ratio is fixed $(\sim1sec, \sim10^{11}\,^{\circ}K)$

Before this era, protons and neutrons were rapidly changing into each other through the emission and absorption of neutrinos. As the average energy level of the particle drops, this process slows, leaving 7 protons for each neutron. This happens since a neutron can change into a proton, an electron, and an antineutrino by itself, but the reverse process requires extra energy. The excess of hydrogen over helium is a direct consequence of the excess of protons, since helium requires neutrons but hydrogen does not.

7) Protons and neutrons form nuclei $(\sim100sec, \sim10^{6}\,^{\circ}K)$

At this stage of expansion and cooling, protons and neutrons come in close proximity $(\sim<10^{-13}cm)$. Protons and neutrons combine after 3 minutes from the Big Bang and at a temperature of 1 billion $^{\circ}K$ to form nuclei of H_e and several other elements. Hydrogen, helium, a few lithium, and deuterium nuclei are formed as strong nuclear forces come into play, confined to such short distances. This nucleo-synthesis process sets the stage for the formation of atoms, stars, and galaxies.

8) Matter nuclei dominate over radiation $(\sim10,000yr.)$

Further cooling and expansion creates more matter from high-energy radiation. Matter loses less energy than radiation. Eventually, matter energy density in newly-formed nuclei dominates the radiation energy density - mainly in mass-less photons. At the end of this process, photons scatter much more with each other than they do with matter. As the exchange between matter and radiation becomes less efficient, photons thermalize and start behaving as thermal blackbody radiation, which can be measured even today as background radiation. Today, after 13.7 billion years, this cosmic radiation glows at 2.728K.

9) Protons and electrons form atoms   (~500,000yr., ~3000°K)

When the temperature drops, the average speed of an electron is not high enough to escape capture by a proton. Atoms start to form. The first neutral atoms form from the nuclei of hydrogen, helium, and lithium. Electrons combine with nuclei to form the first neutral atoms with proper binding energy after 380,000 years. The Universe has cooled to 3000K, and it becomes transparent to light, thereby, emitting cosmic background radiation.

10) First stars form   (~200million-1billionyr., ~20°K)

The temperature of the Universe drops to 20K after 200 million to 1 billion years... As the radiation cools off and de-couples from matter, and almost all the electrons are tied to nuclei in hydrogen, helium, and lithium, then gravitational forces become important. Gravity pulls hydrogen atoms until they collapse and ignite through hydrogen fusion to form the first stars. Small fluctuations in the matter density and gravitational field begin to grow and coalesce. Galaxies begin to form, resulting in the formation of stars, solar systems, planets, and living matter.

11) Heavier elements form   (~2-13 billion yr., 2.728°K)

Stars consume hydrogen and create heavier elements through the fusion process. Thus, elements, heavier than lithium, are produced inside the stars over the lifespan of a star (2 to 13 billion years). These heavier elements are ejected from inside the stars at the end of their lifespan, as they explode into supernovas before settling into old age as white dwarfs, neutron stars, or black holes. Thus, the birth and death of stars led to the formation of solar systems, planets, and living matter.

12) Future of the Universe   (billions of years from now)

Stars will eventually burn all of their fuel supply, e.g., hydrogen, and things in this Universe could get cold and boring.

The Big Bang theory is now almost universally accepted. Penrose and Hawking, in a 1970 paper, using relativity theory, proved that the Universe started with a Big Bang singularity. According to Hawking, however, the assumption of the singularity at the beginning disappears if we include quantum effects.

- Arranging Elements in the Periodic Table

On gaining the understanding of the structure of atoms for various elements, all the known elements are arranged in the form of a neat table - called the Periodic Table, shown in Table. 2.1. All the elements shown in the periodic table have their own unique properties, which find many applications.

The integer that we find in each box of the periodic chart is the number of protons in the nucleus of each atom, which is called the atomic number. The number of protons defines the type of element. If an atom has six protons, it is carbon. If it has 92 protons, it is uranium. The number of neutrons in the nucleus of an element can be different. For example, common carbon 12 has six protons (otherwise it would not be carbon) and six neutrons. However, carbon 13, an isotope of carbon, has six protons and seven neutrons, and carbon 14 has six protons and eight neutrons. Carbon 14 is a radioactive isotope of carbon, because it has the wrong ratio of neutrons to protons, and is unstable.

H (90%), He ((9%), Li was formed 3 min after Big Bang, and B-C-N-O -- Fe formed by fusion in Red Giant stars. Rest of the elements were formed in supernova explosions. Various elements can form different inorganic and organic compounds. The valence electrons in the outermost orbits of their atoms form different types of bonds. One of the simplest and extremely important inorganic compounds is the water molecule, H_2O, formed from two atoms of hydrogen and one atom of oxygen. Organic compounds, based on the carbon element and related chemistry, are also extremely important, since most of the components of our body and organic life are comprised of such compounds. In fact, we owe our existence to this chemistry.

Table 2.1 – Periodic Table of Elements

1	2	3	4	5	6	7	8	9	10	11	12	13	14	15	16	17	18
H 1																	He 2
Li 3	Be 4											B 5	C 6	N 7	O 8	F 9	Ne 10
Na 11	Mg 12											Al 13	Si 14	P 15	S 16	Cl 17	Ar 18
K 19	Ca 20	Sc 21	Ti 22	V 23	Cr 24	Mn 25	Fe 26	Co 27	Ni 28	Cu 29	Zn 30	Ga 31	Ge 32	As 33	Se 34	Br 35	Kr 36
Rb 37	Sr 38	Y 39	Zr 40	Nb 41	Mo 42	Tc 43	Ru 44	Rh 45	Pd 46	Ag 47	Cd 48	In 49	Sn 50	Sb 51	Te 52	I 53	Xe 54
Cs 55	Ba 56	Lu 71	Hf 72	Ta 73	W 74	Re 75	Os 76	Ir 77	Pt 78	Au 79	Hg 80	Tl 81	Pb 82	Bi 83	Po 84	At 85	Rn 86
Fr 87	Ra 88	Lr 103	Db 104	Jl 105	Rf 106	Bh 107	Hn 108	Mt 109	110	111	112	113	114	115	116	117	118
119	120																

f														
La 57	Ce 58	Pr 59	Nd 60	Pm 61	Sm 62	Eu 63	Gd 64	Tb 65	Dy 66	Ho 67	Er 68	Tm 69	Yb 70	
Ac 89	Th 90	Pa 91	U 92	Np 93	Pu 94	Am 95	Cm 96	Bk 97	Cf 98	Es 99	Fm 100	Md 101	No 102	

- Misconceptions about the Big Bang

The following five points address some of the common misconceptions about the Big Bang. It is important to remember the following points:

1) The Big Bang was not an explosion in space, but an explosion of space.

2) The Big Bang did not occur or spread out from a particular location. It occurred everywhere at once.

3) The Universe is self-contained, and it does not need a center to expand away from or into the empty space.

4) Space is dynamic and it keeps expanding faster and faster, with everything receding from each other.

5) At the time of the Big Bang, the totality of space could be infinite Observed from different locations, the early Universe could appear as a pile of overlapping Planck spaces (chunks of 10^{-35}m) stretching infinitely in all directions.

We have uncovered a truly fantastic story during the past century, from the Big Bang to the present. To sum up, hydrogen and helium were formed three minutes after the Big Bang. Starting from helium, the other elements - carbon, oxygen, and all the way up to iron - were formed through the process of fusion. The fusion in heavier stars and subsequent supernova explosions led to the formation of the rest of the heavier elements (~1%) in the Universe.

The air we breathe and the carbon in our body came from the fusion in heavier stars billions of years ago. The chemical reactions occur at much lower temperatures, and involve only the valence electrons and not the nuclei in chemistry. Such chemical reactions account for the formation of the rest of the inorganic and organic molecules and compounds. Thus, everything we breathe, smell, taste, touch, see, or use, came from the elements made inside the stars billions of years ago. As Carl Sagan said, we are literally star-stuff!

2.5 Black Holes and Stars

Let us visit stars and black holes more closely. The formation of stars and black holes, as predicted by Einstein's theories, is a very interesting and important phenomenon. Normally, the gravitational attraction of gas molecules in stars is balanced with the pressure inside the stars from the nuclear fusion reactions... A star gets old as it heats up and fuses its hydrogen into helium. When the star runs out of hydrogen in its core, the core contracts; it heats up and starts fusing the helium into heavier elements, such as oxygen and carbon.

If the initial mass of a star is at least eight times the mass of the Sun, then the fusion process continues until it produces iron. After that, the fusion process stops, since fusing iron does not release but uses up energy. The stars start to cool off and contract in the absence of any useful fuel and fusion pressure, and the matter particles get very close to each other.

However, Pauli's Exclusion Principle does not let matter particles occupy the same space. As matter particles move away from each other with different velocities, the star expands. Chandrasekhar pointed out that the repulsion pressure thus generated is limited, because the maximum difference in the speeds of the matter particles cannot exceed the speed of light. In other words, for massive stars, the gravitational attraction would eventually exceed the limited repulsive pressure.

A star can have three fates depending on its mass. For a star with a mass less than the Chandrasekhar's limit (~1.4 times the mass of Sun), as the star contracts due to gravitational attraction, it achieves a balance with the repulsive pressure and stops contracting. One of the following possible destinies awaits such a star:

White Dwarf – It is a star with a radius of several thousand km and a density of several thousand kg per cm^3. The repulsion pressure of the electrons in the heavy atoms inside the core supports it.

Neutron star – It is a star with a radius of about 15 km, but a density of hundreds of billions kg per cm3. It is supported by the fermionic repulsion pressure of the neutrons in the nuclei, instead of the electrons of the heavy atoms in the core.

Stars that are more massive than the Chandrasekhar's limit might eject their excess mass (supernova) as they shrink and reduce their mass below the limit, thus ending up in one of the above two graveyards. If they do not get rid of the excess mass and are two to three times as massive as the Sun, then they implode inward. This is because the gravitational attraction of these massive stars overwhelms the repulsive pressure of electrons or neutrons. None of the fermionic repulsion pressure is strong enough to prevent the ultimate gravitational collapse into a black hole. Thus, the only stable destiny in such a case is a black hole. Stars born with a mass that's 20-30 times the mass of the Sun usually end up as black holes.

Black hole - A black hole is an object whose gravitational field is so strong that even light cannot escape. Nothing that goes inside it can ever get back out again, not even light. According to current theories, the destiny of all matter that falls into a black hole is crushed to a point of zero volume and infinite density—a singularity.

- Black Holes – What are they?

Black holes are extremely fascinating objects. For a long time, black holes were in the realm of science fiction, but now they have been detected in every galaxy we have observed. Several questions arise about black holes, namely, their true nature, existence, and their role in unraveling the deepest secrets of space, time, and matter. Are black holes relativistic masses inside quantum-sized objects? Do they have anything to do with the mysterious bursts of energy - explosions more powerful than the Big Bang - which astronomers have recently discovered in the far reaches of the Universe?

As stated, a black hole is formed when a very massive star contracts and reaches the end of its life. Gamma-ray bursts are usually associated with the creation of black holes. It is a certain type of cosmic explosion that becomes the brightest thing in the Universe, emitting for a few seconds as much radiation as a million galaxies.

Do not bother looking for one in the sky, though, since most of the light is in the gamma-ray part of the spectrum, a realm that we cannot see. Astronomers observe these gamma-ray bursts (GRB) with space-based telescopes. They generally agree that only the birth of a black hole could supply enough spark for one of these intense flashes, but there remains a great deal of uncertainty over what converts the newborn black hole's energy into the radiation that astronomers detect.

Recent research indicates that a star's magnetic fields are compressed and amplified when the star collapses to a black hole or a highly magnetized neutron star, called a magnetar. Models predict that the fields are strongest — roughly a million billion times that of Earth's magnetic field — along the rotational axis, where they spiral out like an ever-widening corkscrew, according to Giannios at the Planck Institute.

Since magnetic fields have no mass, they are much easier to accelerate than matter. The fields would therefore be more efficient at carrying energy out of the central engine. Outward-moving magnetic fields would eventually dissipate their energy into gamma rays — most likely in a process similar to what happens in a solar flare, according to Erin McMahon at the University of Texas at Austin.

Returning to the process of the formation of a black hole - while a massive star is contracting, the gravitational field at the shrinking surface keeps getting stronger, thus bending the emitted light from the star inwards towards the surface. Its core, which was previously held up by the (fermionic) repulsive pressure, finally collapses due to overwhelming gravitational attraction. The gravitational field, which can bend light, becomes so strong that light bends inwards to the extent that it cannot escape. Nothing else can escape either, since nothing can travel faster than light.

Thus, we have a region of space-time from which it is impossible to escape and reach an observer outside this region. Such a space-time region is called a black hole and its boundary the event horizon7. The path of light rays that fail to escape from the black hole defines the event horizon. Since light cannot escape from it, we cannot see it.

Wheeler, in 1969, gave it the name 'black hole'. Although a black hole cannot be seen, clouds of nearby swirling gases heat up due to violent collisions and the rubbing of particles, giving rise to a temperature of over a million degrees. These quasars radiate and glow, which can be seen near the horizon of the black hole.

It is interesting to trace the scientific evolution of black holes. Indian physicist, S. Chandrasekhar, winner of a Nobel Prize, while attempting to carry forward his work, considered the fate of stars more massive than his limit for white dwarf and neutron stars. On doing so, he came up with the startling conclusion that such a star could collapse to a point.

Chandrasekhar's work along these lines was criticized by scientists, including Einstein, who said in a paper that stars would not shrink to a point of zero volume. Based on general relativity, Oppenheimer, in 1939, showed that a star more massive than Chandrasekhar's limit after collapse would become invisible. Penrose and Hawking showed that a singularity of infinite density and infinite space-time curvature should exist within a black hole

A black hole is a region of space-time where the gravitational forces are so strong that even general relativity and the gravitational theory of Einstein break down. All the matter in the core is crushed out of existence at the singularity. A black hole contains a singularity, which denotes a point where the curvature of space-time is infinite. In simple words, it destroys the fabric of space-time and all the laws of known physics fall apart. However, the laws of physics outside the black hole are not affected. Thus, singularity in a black hole is hidden from the outside by the black hole's event horizon.

An observer or matter falling into a black hole would disappear forever. A single step forward would take one into an abyss without any hope of returning. The fabric of space and time is distorted and unrecognizable. In a black hole, time appears to stand still. Indeed, black holes are extreme geometrical objects, which defy description in classical and quantum physics.

Based on certain solutions of general relativity equations, some scientists believe that matter falling into a black hole reappears in another region of the Universe connected through a wormhole. These solutions, which are most probably unstable, suggest that an observer could travel through space and time through a wormhole if he could avoid hitting the singularity in the black hole.

Thus, one could travel into the past, which is highly unlikely as one could change the events leading to the present state of history. It seems that one would not be able to see the singularity until it hits him. Singularities should either lie in the future, like the black hole due to gravitational collapse, or in the past, like the Big Bang.

General relativity also shows that, under certain reasonable assumptions, an expanding Universe like ours must have begun about 14 billion years ago as a singularity. The currently estimated age of the Universe is several times the life span of an average star. This implies that a large number of stars, bigger than twice the mass of our Sun, must have burned their hydrogen and collapsed, since the Universe began. If the theory is correct, the Universe must contain many black holes.

The gravitational collapse of the center of a large cluster of stars can also lead to the formation of a black hole. These types of black holes can be much more massive than our Sun. Very high-energy particles would be generated near such black holes by the huge amount of matter falling in. Based on recent observations, a black hole seems to exist in the center of every galaxy, including our galaxy, the Milky Way.

- Do black holes really exist?

How did they confirm the existence of black holes? Some time ago, a black hole, 300 million times more massive than our Sun, was discovered in the galaxy called NGC 7052 (Fig. 2.6)..

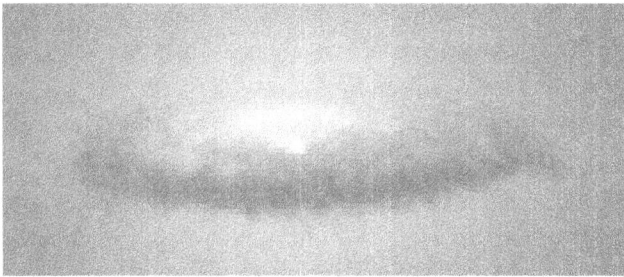

Fig. 2.6 - Black hole in galaxy NGC 7052 - by Hubble Telescope

Since light cannot escape from a black hole, no photons are emitted and we cannot observe black holes directly. However, the Dressler and Nuker team came up with the following idea in 1983: We should look for evidence of its enormous gravitational force and its pull on stars; we should measure how fast the stars are moving as they get near a black hole. They tracked the movement with the help of a spectroscope and looked for a sudden shift in the position of a vertical dark band. They observed such a shift in the center of nearby galaxy, Andromeda

In 1973, theorists came up with the idea that magnetic fields could drive the generation of light by gas falling onto black holes. Thirt years later, using Chandra, Miller and his team provided crucial evidence confirming the role of magnetic forces in the black hole accretion process. Chandra Program is manged by NASA's Marshall Space Flight Center, Huntsville, Ala., for the agency's Science Mission Directorate. The Smithsonian Astrophysical Observatory controls science and flight operations from the Chandra X-ray Center, Cambridge, Mass.

The X-ray spectrum, the number of X-rays at different energies, provided by Chandra X-Ray Observatory, showed that the speed and density of the wind from J1655's disk correspond to computer simulation predictions for magnetically-driven winds. The spectral fingerprint also ruled out the two other major competing theories to winds driven by magnetic fields.. This deeper understanding of how black holes accrete matter also explains other properties of black holes, including how they grow. According to Danny Steegh at the Harvard-Smithsonian Center for Astrophysics, this understanding, in addition to accretion disks around black holes, might also lead to understanding the role of magnetic fields in disks detected around young sun-like stars where planets are forming, and objects called neutron stars.

Astronomers using NASA's Chandra X-ray Observatory also found evidence of the youngest black hole known to exist in earth's neighborhood. This object is a remnant of SN 1979C, a supernova in the galaxy M100 approximately 50 million light years from Earth. The light from the supernova took 50 million years to reach Chandra, and we see the object as it appeared when it was just 30-years-old old baby. Thus, this black hole provides a unique opportunity to watch this type of object develop from infancy. The black hole could help scientists better understand how massive stars explode, and how they leave behind black holes or neutron stars.

The UCLA scientists also discovered a black hole in the center of our own galaxy, the Milky Way. Surprisingly enough, a black hole has been discovered near the center of every observed galaxy. Observations for a large number of galaxies by the upgraded Hubble Telescope recently seem to confirm that every galaxy, active or inactive, has a black hole at its center. These black holes, some very massive and others not so massive, are feeding on stars as the stars come near and are sucked in.

It is estimated that about ½% of the entire mass of each galaxy is due to the black hole at its center. In 2007, scientists pinpointed the precise locations of a pair of super massive black holes at the centers of two colliding galaxies, 300 million light years away. Infrared images made by the Keck II telescope in Hawaii reveal that the two black holes at the center of the galaxy merger known as NGC 6240 are each surrounded by a rotating disk of stars and cloudy stellar nurseries.

In 2007, the astronomers discovered a stellar black hole, which is much more massive than theory predicts. The astronomers were puzzled. As stated, stellar black holes form when stars with masses around 20 times that of the Sun collapse under the weight of their own gravity at the ends of their lives. Most stellar black holes weigh in at around 10 solar masses when the smoke blows away. The computer models of star evolution have difficulty producing black holes more massive than this. The newly-weighed black hole is 16 solar masses. It orbits a companion star in the spiral galaxy, Messier 33, located 2.7 million light years from Earth. Together, they make up the system known as M33 X-7.

The world's heaviest black hole was discovered in 2008 by a team in Finland. The monster celestial object is 18 billion times more massive than our own Sun - six times larger than the previous record. The object, called OJ287, is orbited by a smaller black hole, which allowed its mass to be measured very accurately. The finding also enabled the researchers to test Einstein's theory of gravity for the first time in a strong gravitational field. The binary black hole system powers a quasar - a compact halo of matter, which radiates enormous amounts of energy. Details of the finding were presented at the 211th meeting of the American Astronomical Society (AAS) in Austin, Texas.

The most distant explosion was recorded in the fall of 2005, which indicated the birth of a black hole near the beginning of time. The event was noted by NASA's orbiting Swift observatory as a gamma ray burst, which lasted more than 8 minutes. An afterglow in visible light, and other wavelengths, followed this burst of high-energy radiation on the electromagnetic spectrum.

The burst, named GRB 050904, originated 12.8 billion light years away, which means that it occurred 12.8 billion years ago, as the light took that long to reach us. It erupted very nearly at the beginning of time, since our Universe is about 13.7 billion years old. It was more chaotic and lasted longer than expected. A strange sequence of events occurred, as the explosion of a massive star first settled down, but then fired back up several times toward the end. Astronomers speculate that the black hole did not form instantly, as theory predicts, but that it was a prolonged process.

A surprising behavior has also been observed for stars, which are far away from for the black hole at the center of the galaxy. It concerns the speed of stars circling at the outer edges of a galaxy, and its relation to the mass of the black hole. As stated earlier, it has been a common belief that, after the Big Bang, as clouds of hydrogen and helium gases swirled around, somehow the lumpiness, stars, and galaxies formed. One would normally expect that, at the edge of a galaxy, the stars would be far from the center of the galaxy, where the black hole resides. Thus, the presence of the black hole – no matter how massive it is - will have no significant effect on the speed of the star, a property called 'sigma' by cosmologists.

The mass of black holes for various galaxies was plotted against the speed of the stars at the edge of these galaxies to confirm this. Scientists were, however, surprised to find a linear relation, namely, the speed increased linearly with the mass of the black hole. Reese and Silk had earlier developed some physics for it. Silk, after rethinking through the old ideas, concluded that these stars at the outer edges were linked to the black hole at the birth of the galaxy. In other words, black holes are not just destructive monsters in the Universe, but they may be responsible for the formation of the galaxies.

A hypothesis based on this idea is as follows. As the clouds of gases were swirling after the Big Bang, they were pulled together and collapsed to form giant black holes. As black holes were feeding on these gases, creating quasars, it caused temperature changes in the gas clouds, which led to the formation of stars and galaxies. Perhaps, the conditions in the early Universe and the irregularities required for the formation of stars and galaxies might have created 'primordial' black holes in the Universe. In fact, one string theory model for a pre-bang scenario even suggests that the interior of an unusually large black hole might have become our Universe.

- Black Hole & Thermodynamics

There seems to be a direct correspondence between the properties of a classical thermodynamic system and the properties of a black hole. We can substitute black hole gravity at the event horizon for temperature, its mass for energy, and area of the event horizon for entropy in the thermodynamics laws to obtain the laws for black holes. This correspondence becomes obvious when we state the four laws of thermodynamics and for the black holes together, as follows:

- *Thermodynamics*: The temperature of a system in thermal equilibrium has the same value everywhere in the system
- *Black holes*: The surface gravity at the event horizon is constant: it has the same value everywhere on the event horizon.
- *Thermodynamics*: The change in energy of a system is proportional to the temperature times the change in entropy
- *Black holes*: The change in mass of a black hole is proportional to the surface gravity times the change in area.
- *Thermodynamics*: The total entropy of a system can only increase.
- *Black holes*: The surface area of the event horizon can only increase.
- *Thermodynamics*: It is impossible to lower the temperature of a system to zero through any physical process
- *Black holes*: It is impossible to lower the surface gravity at the event horizon to zero through any physical process.

According to the second law, it is impossible for black holes to decay and disappear since a black hole cannot get smaller or split into a smaller black hole. This property, along with the first property - namely, the surface gravity - is constant over each event horizon; these properties are obtained by solving the Einstein equations, which do not include quantum mechanics. In the classical black hole theory, the event horizon is an absolute barrier and nothing can get out of the black hole.

- Black Hole Information Loss Dilemma

According to Hawking, black holes also have the information loss problem, which is quite perplexing. Scientists say that when a body collapses to form a black hole, it loses a large amount of information. It takes a large number of parameters to describe a collapsing body. However, the black hole that forms is completely independent of the nature of the body or type of matter that collapsed. After the formation of a black hole, we can possibly measure only its mass and rate of rotation and nothing else. In classical theory, this loss of information would not really matter.

One might argue that all the information about the collapsing body was still inside the black hole. The observer would never actually lose sight of the collapsing body. However, it would appear to slow and get dim as it approached the event horizon. Nevertheless, the observer could still see what it was made of and how the mass was distributed.

Hawking said that quantum mechanics and the uncertainty principle added complexity to the simple picture. First, regarding the loss of information, a collapsing body before it crosses the event horizon can send out only a limited number of photons, which are not sufficient to carry all the information about the collapsing body. This means that, in quantum theory, an outside observer cannot measure the state of the collapsed body. One might say that this it is not so important, since the information is still inside the black hole, even if we could not measure it from the outside.

It is known that in a vacuum, the quantum fluctuations create virtual particle-antiparticle pairs that annihilate each other before they can be observed. These particles are called virtual particles, since we cannot observe them with a detector, unlike real particles. However, we can indirectly measure their effects, which agree with theoretical predictions. The virtual particle has positive energy and the antiparticle in the pair has an equal amount of negative energy to make the energy sum zero, since energy cannot be created out of 'nothing'.

Hawking argued that an antiparticle with negative energy at the horizon might be sucked into the black hole, due to the strong gravitational field inside the black hole virtual, forgetting and leaving the positive energy particle outside the black hole. This results in negative energy flowing into the black hole, reducing its mass, and an equal amount of positive energy (equivalent to the reduced mass of the black hole) flowing or radiating away from the black hole.

Thus, according to Hawking, quantum theory causes black holes to radiate and lose mass. As a black hole loses mass, its temperature and the rate of emission increase, thereby losing mass at a faster rate. It seems that they will eventually disappear completely, taking with them the information inside of them. Hawking gave arguments in the 1970s that proved that this information is lost and does not come back in some form.

This loss of information would have introduced a new level of uncertainty into physics, over and above the usual uncertainty associated with quantum theory. By the way, a black hole, at least twice as massive as the Sun, would have a temperature of $\sim 10^{-7}$ degrees above absolute zero, compared to the microwave radiation at 2.70 above absolute zero. Black holes that are more massive would have an even lower temperature. Thus, such black holes would absorb more radiation than they would emit.

Hawking has gone back and forth about the information inside a black hole. At one time, he held that the 'information' falling into the black hole is lost forever, but recently, he said that the contents of a black hole would leak out in the form of 'Hawking radiation', until the black hole itself dissipates. Hawking suggested that black holes constantly radiate energy and lose information, because of the activity in the empty space near the event horizon, which is bubbling with virtual particles.

Hawking has suggested that hidden information in the black hole eventually evaporates. However, Hawking admitted in 2005 that he was wrong in his calculations. According to his new calculations, the radiation flowing out of a black hole does carry information about the black hole out to the Universe, but it is in mangled form. He said that the information is not lost, but it is not returned in a useful form.

"It is like burning an encyclopedia. Information is not lost, but it is very hard to read."

At present, we do not have a quantum gravity theory that combines Einstein's general relativity or gravitational theory with quantum mechanics. However, using a sleight of hand, we can combine quantum mechanics with classical general relativity by looking at particle scatterings in curved space-time, where the space-time curvature cannot react to the scattering quantum particles. This is not quite correct, since we use quantum physics for the particles, keeping the gravity part classical. Using such an approach,

Hawking obtained the above-mentioned strange results, which he modified recently. The definition of particle and antiparticle depends on the observer, which is against the usual rules of the theory of general relativity. The implication of this is that the number of particles being counted depends also on the observer. The much-awaited quantum gravity theory might add to our understanding of black holes and our Universe.

In a 2008 report in Nature, US researchers claimed that they have worked out how black holes emit jet streams of particles at close to light speed. They said that the streams originate in the magnetic field near the edge of the black hole. According to them, the jets accelerate and focus within this region. Scientists have also long wondered how stars develop in such extreme conditions. Normally, the molecular clouds - the normal birthplace of stars - would be ripped apart by the immense gravity.

In August 2008, astronomers claimed to have found out how stars can form around a massive black hole, defying conventional wisdom. The researchers said that stars can form from elliptical discs - the relics of giant gas clouds torn apart by encounters with black holes. The finding was based on computer simulations of giant gas clouds being sucked into black holes. According to Ian Bonnell from St. Andrews University, UK, these simulations show that young stars can form in the neighborhood of super-massive black holes, as long as there is a reasonable supply of massive clouds of gas from further out in the galaxy.

To sum up, black holes are fascinating, and so is its physics. Despite substantial progress, however, we still do not have a complete understanding of black holes, their role, and their physics.

2.6 Structure of the Universe

Recent evidence confirms the expansion of space. Scientists also believe that the Universe is homogeneous, isotropic, and has a flat structure. Let us visit these concepts and the related evidence.

- An Expanding Universe & Measuring the Expansion Rate

To understand the evidence regarding the expansion of the Universe, we need to understand the concept of redshift. Redshift is the shift of the light. spectrum toward red, and it is an indication that the space is expanding. It is believed to occur when space is expanding like a balloon.. as illustrated in Fig. 2.7. The light traveling across the expanding space from a distant galaxy is stretched, its wavelength increases, and its spectrum shifts toward red,

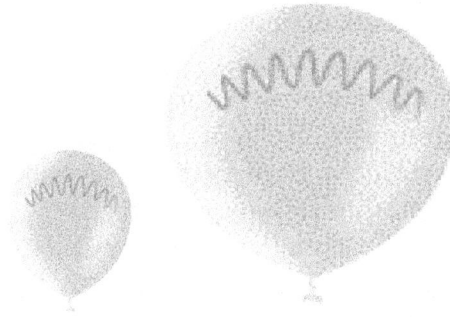

Fig. 2.7. Space expansion like a balloon

Scientists differ on their views about the origin of redshift. For example, in the approach based on the Doppler effect, for a given redshift (z), the speed of recession (V) is obtained by using the simple equation, $z = c/(c-V)$, where c is the speed of light. According to the Einstein-De Sitter model, the Doppler effect does not cause the redshift, due to the galaxies' speed of recession. The redshift, according to this model, is caused by the space dilation in which the galaxies take part, without moving away with respect to space itself. The redshift, thus, originates because of the stretching of the light wave, due to the dilation of space8.

The higher the redshift, the more distant the object is that emitted the light. Light from very distant galaxies takes billions of years to travel to Earth, giving a snapshot of the Universe at that instant long ago. The basic idea is to measure the cosmic expansion of space by comparing the distances of far-off objects with their redshifts. According to the Supernova Cosmology Project, distant supernovae tell an interesting story regarding the acceleration of the expansion rate.

This Supernova Cosmology Project uses powerful and sensitive instruments, which include giant telescopes on the ground, the Hubble

Space Telescope, charge-coupled devices instead of photographic plates, and supercomputers. The project uses Type Ia supernovae as standard candles, which are exploding stars as bright as entire galaxies that can be seen across billions of light years.

These thermonuclear explosions emit most of their energy in a few weeks. During that time, each gives off nearly the same amount of light. At high redshifts, supernovae in distant galaxies are dimmer than they would be if expansion were slowing. Thus, they must be located farther away than expected for a given redshift. This evidence suggests that the expansion rate of the Universe is accelerating.

We regard Type 1a supernovae as uniform in brightness. We calculate the distances on how bright they appear to be through telescopes. According to Nugent, Type 1a supernovae are reliable distance indicators because they have a standard amount of fuel - the carbon and oxygen in a white dwarf star - and they have a uniform trigger. They form when a white dwarf pulls enough matter from a nearby companion star to explode in a violent thermonuclear reaction.

According to Chandrasekhar, no white dwarf can be more massive than about 1.4 solar masses before it self-destructs. However, an object, known as SNLS-03D3bb, discovered in April 2003, and classified as a Type 1a supernova, appears to contravene this rule. This supernova is twice the brightness expected, which suggests that it arose from a star much too massive to exist. The fact, that SNLS-03D3bb is well over such a mass, opens up a Pandora's Box.

Scientists said, in May 2007, that they have detected the brightest stellar explosion ever recorded, a new breed of supernova that may well be repeated sooner than they previously thought. The ground-based telescopes, as well as NASA's orbiting Chandra X-Ray Observatory, observed this violent explosion in a galaxy far from our own Milky Way. The observations also hint that an erupting star in our own galaxy, called Eta Carinae, could be close to the same kind of blast, according to astronomers.

The Hubble's constant, Ho, is also an important element because it allows us to determine the distances to more distant objects in the Universe. Hubble's law quantifies the expansion rate of space. It states that the recession velocity (v) of a galaxy away from any observer is directly proportional to its distance (d) from the observer, v= Ho.d. The recession velocity is the rate at which the space is stretching--not just around us but also around any observer. Thus, if we know v and Ho, we can determine the distance (d). Einstein's equation also leads us to the Hubble parameter, H (t)3.

More precisely, we can measure the distance, if we can find out more about the mysterious dark energy that's responsible for this expansion of the Universe. As stated, astronomers have used variable stars and a special kind of supernova to make their distance estimates. They are now using two new measuring sticks. One of the yardsticks is a big and bright type of variable star, known as an ultra-long-period Cepheid variable, or ULP for short. The other yardstick makes measurements using the radio emissions from the super-massive black holes lying at the center of many galaxies, including our own.

The fine-tuning of these measuring techniques in the years ahead can give us a more precise measurement of the Universe's expansion rate over time, tied to the Hubble constant. Right now, the estimate is based on a combination of distance scales, with short-period Cepheid variables used for relatively close measurements and Type 1A supernovae for farther measurements.

WMAP study estimates Ho = 71 km/sec/Mpc (with a margin of error of about 5%) or $\sim 2.31 \times 10^{-18}$ sec-1. Reciprocal of the Hubble constant, Ho, gives the age of the observable Universe as 4.32×10^{17} sec or 13.7 billion light years. In cosmology, since nothing can travel faster than the speed of light in space, our observable Universe is limited to ~13.7 billion light years. Galaxies beyond a certain distance (Hubble distance, c/H = ~13.7 billion light years), according to Hubble's law, must be receding at speeds higher than the speed of light. This is possible, since the recession velocity is caused by the expansion of space and not motion through space, which would be subject to the special relativistic limit.

Next, we visit the geometry of the Universe. The above discussion would be valid, if the spatial geometry of the Universe were flat and not curved. If the Universe were open with negative curvature, distant supernovae might appear deceptively fainter, mimicking acceleration. It is also possible that some of the intergalactic dust that absorbs their light could distort the observations of supernovae. The Supernova Cosmology

Project will also study the curvature of the Universe and detect possible distortions. We will discuss recent independent evidence, which suggests that the spatial geometry of the Universe is flat.

- Homogeneous, Flat-Space Isotropic Universe

One of the important assumptions behind the Big Bang theory is that space is isotropic and homogeneous. In other words, at every time, space looks the same in every direction (isotropic) and at every point (homogeneous). In cosmology, we call such a space, 'maximally symmetric', which is a reasonable assumption at large distances in cosmology. What exactly do we mean by the 'curvature of space'?

We probably understand what a curvature of an object in space means. If we could look at a very large spherical ball all at once, we could immediately see the curved surface of this large sphere. However, if an ant is crawling on this surface, it cannot see the whole sphere all at once. The ant might think it was crawling on a large, flat, two-dimensional plane, and could find out the truth only if it could carefully measure distances on the sphere.

Einstein's theory of general relativity unified space and time into a single geometric entity, called space-time. Space-time has geometry just like ball, and it could curve, just like the surface of the ball. Thus, we associate curvature with space-time. However, to determine the true curvature of space geometry, we must be able to describe the whole of space and the whole of time, i.e., everywhere and forever at once.

In other words, description of space-time geometry implies including everything that has ever happened and will ever happen in space-time. This would be impossible to know. However, scientists are ingenious, and they make simplifying but realistic assumptions to come up with models that can approximate our Universe reasonably well at very large distances.

The first assumption made is that space-time can be separated into space and time. We know that this is certainly not the case below the Planck length scale. It is also not true near the black hole, where the space-time curvature is twisted, and cannot separate. Except for these anomalies, we can separate time and space in our Universe. The second assumption scientists make is that space is isotropic and homogeneous, i.e. maximally symmetric.

To solve the Einstein equation of general relativity for the evolution of space-time geometry, scientists consider three basic types of energy that could curve space-time. These energies are radiation, matter, and vacuum energy. The radiation and matter energy are lumped together and treated like uniform gas with pressure relating its density as the equation of state. The Einstein equation under these assumptions reduces to two ordinary differential equations. The solution gives us the space geometry and changes in space size with time9.

The assumption of maximal symmetry for space implies that the fabric of space must have a constant curvature at large (except for distortion near huge material objects). Otherwise, it would look different from every other point in the direction where the curvature was different.

Constant curvature narrows down the options for the geometry of space to three - namely, positive, negative or zero curvature. Fig. 2.8 illustrates these three structures.

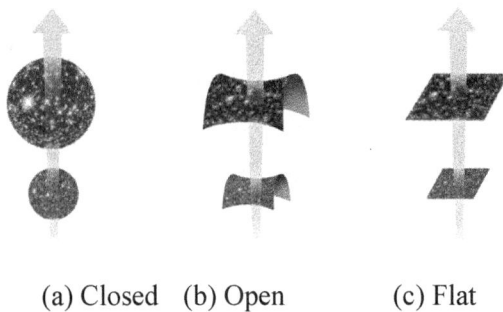

(a) Closed   (b) Open   (c) Flat

Fig. 2.8 – Possible space structures of the expanding Universe

One can visualize multi-dimensional space with a positive constant curvature by thinking of a sphere, which leads to a closed Universe (Fig. 2.8 a NASA). In such a Universe, space expands from zero volume in a Big Bang to a maximum volume and then starts contracting back to zero volume in a big crunch.

The negative constant curvature of N-dimensional space can be visualized by thinking of a hyperboloid – a two dimensional pseudo-sphere. Space in such a Universe has infinite volume, expanding forever, and representing an open Universe (Fig. 2.8 b NASA). For zero constant curvature, space is obviously flat in every direction. It extends infinitely in every direction, representing an open Universe that is expanding forever (Fig. 2.8 c - NASA).

Closed, open, and flat universes can re-collapse, expand forever, rip apart, or reach a tenuous balance. The invisible dark energy could propel even a closed Universe to eternal expansion.

If we ignore the vacuum energy and consider matter and radiation only, the curvature of space also tells us about the evolution of space with time. The determination of the evolution of space in time from the Einstein equation becomes difficult when we include vacuum energy, since energy conservation and the equations of state (pressure-density relations) determine the energy density changes as space evolves in time.

Several different observations suggest that space is not only expanding at an accelerated rate, but it is also flat and not curved. Current observations indicate that our Universe seems to have enough energy density and the cosmological constant to provide critical density and zero spatial curvature. The measurements of cosmic microwave background radiation indicate that the Universe space is flat. By far the most successful explanation for the flatness of the Universe space is the inflation theory.

Another interesting observation regarding the structure of the Universe is the cosmic web. A cosmic web is woven from a hierarchy of structures. We can observe the Universe starting from a few million light years all the way to 13.7 billion light years away. We observe stars, individual galaxies, groups of galaxies, clusters of galaxies, and finally, super clusters of galaxies, near the farthest point in the Universe. This web structure is believed to be due to the gravitational instability that was discussed with Newton's law of gravity. It implies that gravitational instability results from unbalanced gravitational forces, caused by the slight fluctuations in the density of matter.

2.7 Problems with the Big Bang Model

Why did the Big Bang occur? Where did the energy come from? What happened before the Big Bang? When did time begin? Einstein's theories imply that the Universe was originally a point singularity with infinite density, which is physically not possible. At this stage, science also does not have any conclusive answers to questions about the events before the Big Bang. Martin Rees has written an interesting book, Before the Beginning. However, most questions remain unanswered regarding the pre-bang era.

Before we visit some of the problems concerning the Big Bang model, let us emphasize the following fact. The Big Bang model, despite its shortcomings, is the only satisfactory scientific model that can explain many observed phenomena in this Universe. Science does continue to seek better models and theories that could answer the unanswered questions.

- Before the Big Bang

According to the Big Bang model, our Universe had a beginning. However, according to the principle of causality, an event is caused by an earlier event, and the chain continues. One might ask, 'What caused the first event in the chain?' Nature works in cycles, and it would not be far-fetched to assume that the Universe exists eternally, going through this cycle of cause and effect. What we call the origin of the Universe is caused by a previous effect. Several other related questions also arise. The laws of physics responsible for the creation and evolution of the Universe – when and where were they formulated, especially when there was no space or time before creation?

One also needs to answer the question, 'Where did the energy for the Big Bang come from?' In other words, how could the most revered law of science, the conservation of energy, be violated - assuming the equivalence of mass and energy through Einstein's famous equation: $E = mc^2$?

Some scientists believe that Heisenberg's Uncertainty Principle can resolve this dilemma. We could assume that the Universe was created from a quantum fluctuation, based on the laws of quantum physics. However, we end up with another dilemma - when and where were the laws of quantum physics formulated? Science has so far avoided answering such questions. However, for science to be complete, it must at least attempt to answer such questions.

Hawking says that it might be beyond our powers, but we must at least try to understand the beginning of the Universe based on science. Hawking even discussed the idea of an imaginary time in addition to the real time, which leads to a conclusion that the Universe need not have a beginning or an end. One could avoid specifying boundary conditions if the histories of the Universe in imaginary time are closed surfaces, like the surface of the Earth. One might also speculate that the chain of events is closed and the Universe has no beginning or an end.

Let us briefly visit some of the ideas concerning the problems of the Big Bang and pre-Big Bang era, starting with the energy problem. According to the uncertainty principle, subatomic particles at quantum level do not necessarily have to follow the principle of causality. Matter and antimatter particles can appear, annihilate each other, and disappear in the so-called quantum foam frenzy, due to quantum fluctuations in space at the Planck scale. It is like borrowing and repaying quickly the loan to balance the books of nature on energy before anyone can notice.

How could the Universe evolve from such borrowed energy coming out of nothing for an extremely short moment before it is returned? According to the uncertainty principle, the more energy it borrows, the less time it gets to repay and balance the books. Thus, there would not be enough time for the Universe and life to evolve.

Speculating further, the Universe could use a trick that a government might use to extend or reduce payments, namely, inflation. The Universe could inflate extremely fast from a space quantum at Planck scale to an enormous size at a rate faster than the speed of light. In fact, the inflationary theory proposed by Guth in 1980 developed a scenario for the fast inflation or expansion of the Universe in a fraction of a second.

According to Kistler, the Universe could thus repay, or at least pretend to repay, its debt by offsetting the positive of matter/energy with the negative of gravity. It reminds one of a clever businessperson, who borrows some money initially from the bank, based on his business model. He then builds and expands his business at an extremely fast pace, paying of his original debt, as he borrows new money against expanding business as collateral.

String theory has also proposed different scenarios for the pre-Big Bang era. For example, Veneziano combined T-duality with the symmetry of time reversal in 1991 to propose a pre-Big Bang model. It attempts to paint strange scenarios, like the pre-bang Universe being a mirror image of the post-bang Universe, or proposing a violent transition from acceleration to deceleration, instead of a Big Bang, as the origin of the Universe. It includes the basic idea of inflation theory, which states that the Universe had to undergo acceleration to become homogeneous and isotropic.

String theory also proposes another model, the ekpyrotic model, for the Universe before the Big Bang. This model claims that our Universe is one of many D-branes floating in a higher dimensional space, attracting each other and colliding occasionally. The Big Bang could thus be the impact of a collision of another brane into ours.

- Three Major Problems

The other three major problems concerning the Big Bang model are:
  1) The flatness problem
  2) The horizon problem
  3) The magnetic-monopole problem

For answers to these problems, we probably have to look at events before the Big Bang. As regards the flatness problem, we've already discussed that the observed lumpiness in the temperature of the cosmic microwave background radiation indicates a flat space that is expanding forever. Einstein's equation predicts that even a slight deviation from flatness in an expanding Universe would grow very large. Therefore, initial events must have left the Universe with immeasurably small deviation from flatness. We shall soon see how the inflation theory addresses this problem.

Regarding the horizon problem, observations of the cosmic microwave radiation also show that our Universe is amazingly smooth in all directions. It is filled with highly isotropic thermal radiation with a thermal radiation temperature of 2.73K. However, this radiation can be uniform only if the photons mix around or thermalize through particle collisions. Photons moving at the speed of light cannot get from one side of the Universe to the other in time to account for the observed isotropy in the thermal radiation. Events before the Big Bang left the Universe amazingly smooth in all directions.

Again, inflation theory provides an answer. As regards the monopole problem, every magnet has two poles, but grand unified theories and string theory predict magnetic monopoles to exist. The Big Bang should have produced many such poles that would make one hundred billion times the observed energy density in the Universe. However, we have yet to find one magnetic monopole. Despite wide acceptability of the Big Bang model, it is obvious from the preceding discussion that several questions remain unanswered.

2.8 Inflation Theory and Recent Observations

The inflationary theory, proposed by Guth in 1980, seems to resolve the three major problems. To solve the flatness problem, the inflation model assumes that the Universe went through an inflationary phase before the radiation-dominated era, when only uniform vacuum energy was present. Think of a balloon being inflated rapidly, which flattens out its initial curvature as it expands.

In the inflation model, the Universe goes through a rapid expansion, like inflating a bubble of pure vacuum energy with no matter or radiation present, which leaves the Universe space extremely flat. The inflation scenario thus assumes a special field, called 'inflaton'. This field drives the exponential expansion and lets the Universe roll downhill in potential energy. According to some estimates, this rapid exponential expansion of the false vacuum bubble lasts about 10^{-32} second, and the bubble expands around 70 e-fold, i.e., in de Sitter Model $e^{2H.t} \sim e^{70}$. This gives a vacuum energy density of $\sim 10^{+93}$ J/m3 ($<$ Planck energy density $\sim 10^{+113}$).

Nevertheless, it would take care of both the flatness and horizon problems. This potential energy in the vacuum converts into the kinetic energy of matter and radiation through processes of the particle physics. Then, the standard Big Bang starts. It explains why the Big Bang started with a flat spatial geometry and the Universe is still quite flat[10].

The inflationary model also solves the horizon problem. The vacuum pressure accelerates the expansion of the Universe exponentially. The photons can traverse more space, which does not contain matter. Light also gets farther and faster, because space is expanding faster and it would thus have crossed the existing Universe by the end of the inflationary era. Therefore, the isotropy of radiation from the Big Bang is consistent with the speed of light.

The inflationary theory also solves the monopole problem, since there would be only one magnetic monopole for the vacuum energy bubble, i.e. for the entire Universe. In fact, cosmologists have widely accepted the inflation theory. For example, Turner in a recent paper, "Dark Energy and the New Cosmology", discusses the new standard cosmology – a successor to the standard hot big-bang cosmology. According to him, this new standard cosmology is characterized by:

1) Flat, accelerating Universe
2) Early period of rapid expansion (inflation)
3) Density inhomogeneities produced from quantum fluctuations during inflation
4) Composition: 2/3rds dark energy; 1/3rd dark matter; 1/200th bright stars
5) Matter content: $(29 \pm 4)\%$ cold dark matter; $(4 \pm 1)\%$ baryons; $\sim 0.3\%$ neutrinos

Although not as well-established as the standard hot big-bang model, the evidence is mounting in support of a new model, as discussed in Turner's paper. The existence of acoustic peaks in the CMB power spectrum and the evidence for a nearly scale-invariant spectrum of primeval density perturbations ($n = 1 \pm 0{:}07$) is exactly what inflation predicts (along with a flat Universe).

Recent observations, however, show that the composition of the Universe differs slightly from Turner's numbers. From the observed data by NASA's WMAP, one can estimate the composition of the Universe as 73% dark energy, 23% dark matter, and 4% ordinary matter.

Support for the inflation theory comes from certain observations. These observations paint an interesting picture before the completion of the first fraction of a second of the Universe, when the sub-atomic scale activity and tiny 'quantum fluctuations' drove the Universe towards stars and life. With the sudden expansion of a pinhead-sized portion of the Universe in a fraction of a second, random quantum fluctuations inflated rapidly from the tiny quantum world to a macroscopic landscape of astronomical proportions.

This belief is based on the observation that the microwave afterglow light from the Big Bang has a uniform temperature across the sky. Different parts of the Universe have not had time to come into an equilibrium with each other *unless* the regions had exponentially inflated from a tiny patch. The only way the isotropy could have arisen is if the different regions were in thermal equilibrium with each other early in the history of the Universe, and then rapidly inflated apart.

The inflation theory, though widely accepted, does raise some questions. Just after the Big Bang, why did the Universe blow up at an incredible rate? What caused this blow-up? What disturbed the equilibrium of the inflaton field to create the Universe? What are the initial conditions that produce inflation? How can the inflationary phase be made to last long enough to produce our Universe? In other words, the inflation scenario, invented to circumvent the unexplained initial conditions of the Big Bang, requires contrived initial conditions of its own.

2.9 Six Numbers Decide the Fate of the Universe

In 1999, Martin Rees, in his book, Just Six Numbers, advanced an interesting concept related to our Universe. Rees states that just six numbers, set at the time of the Big Bang, govern this Universe. Thus, according to Rees, these six numbers constitute a 'recipe' for our Universe. The stars, planets, and humans could not exist, if any one of these numbers were different. Two of these numbers are related to the basic forces. The next two fix the size and texture of the Universe and decide if it would continue forever. The remaining two fix the properties of space itself.

The six cosmological numbers, described by Rees, are given in Table. 2.2. The first number D is the number of spatial dimensions. Life, as we know it, could not exist if D were two or four. Time is considered a fourth dimension, but it is different, since it has a built-in arrow, and it always points only towards the future.

The second number, called the cosmic number, Omega (Ω), is a measure of the amount of material in our Universe - galaxies, diffuse gas, and dark matter. It is the ratio between the actual and critical density - required to halt the expansion of the Universe. It indicates the relative importance of gravity and expansion energy in the Universe. According to the inflationary theory of the Big Bang, omega should be 1, but its exact value is yet to be measured. With the recent DASI observations, however, the evidence for flatness is now quite firm, $\Omega = 1.0 \pm 0.04$.

#	Description	Value
D	Space Dimensions	3
Ω	Actual and critical density Ratio of the Universe	1
ε	Fine structure constant, $k.e^2/hc$	1/137
N	Ratio of electrical to gravitational force between two protons	10^{+36}
Q	Fraction of its rest mass energy needed to pull cosmic structure apart	10^{-5}
Λ	Energy of the vacuum - cosmological constant	0.7

Table. 2.2 - Six Cosmological Numbers

If we could dismantle all the visible matter in the Universe and spread it smoothly in space, we would have about one atom in a 10 cubic meter space. This is an extremely small value. Even if we assume that dark matter weighs ten times ordinary matter, it would only contribute to a Ω value of 0.2. However, scientists strongly believe that the value of Ω is 1, and the actual density (ρ) of our Universe equals the critical density, ρc ($=3H^2/8\pi G$), where H is the Hubble parameter whose current value is the Hubble constant (Ho). This is because if $\Omega >> 1$, the Universe would have collapsed long ago, and with $\Omega << 1$, no galaxies could form.

The third number is called the 'fine structure constant number' (also called alpha α)ε. It defines the binding of the atomic nuclei and the formation of atoms on Earth. It is defined by the expression, $\varepsilon = k.e^2/hc = 0.0072972$, where "e" is the charge of the electron and $k = 1/(4\pi\varepsilon o)$, εo – permittivity of the vacuum space. It is also the ratio of the electric field energy, ke^2/R at a distance of Compton radius, R (= h/mc) to the rest energy, mc^2 for an electron.

According to Rees, the value of the constant ε controls the power from the Sun, and the transmutation of hydrogen into all the atoms of the periodic table. Carbon and oxygen are common and gold and uranium are rare, because of what happens in the stars. The epsilon ε had to be 0.0072972 (~1/137) if we were to exist. Measured change over 10 billion years in the fine structure constant ε value is about 1 part in 100,000.

The fourth number N measures the strength of the electrical forces that hold atoms together, divided by the force of gravity between them, i.e., $(k.e_p^2/G.m_p^2 = 1.23 \times 10^{+36})$ where, e_p is proton's charge and m_p is its mass. Since N is so huge, the cosmos is so vast. If it had a few less zeros, only a short-lived and miniature Universe could exist. No creatures would be larger than insects, and there would be no time for evolution to lead to intelligent life.

The fifth number Q denotes the fraction of the total rest mass energy (mc2) of a cosmic structure that is required to pull it apart. It characterizes how strongly a cosmic structure is bound together by gravity. The fabric of our Universe depends on this number. The seeds for all cosmic structures - stars, galaxies, and clusters of galaxies – are all imprinted in the Big Bang. If Q were smaller, the Universe would be inert and structure-less. If Q were much larger, it would be a violent place, dominated by giant black holes.

The sixth number, the cosmological constant – lambda (Λ) - is responsible for the expansion of the Universe. A new force, 'antigravity', vacuum, or dark energy seems to control the expansion of our Universe. We discussed it with dark matter in the Standard Model. It accelerates the expansion of the Universe in time, which one can see in the red shifts of distant supernovae. If Λ were not very small, cosmic evolution could not take place and galaxies and stars could not form. It was measured in 1998, but its precise value is still uncertain.

A question arises whether these six numbers are (and have always remained) constant since the Universe started expanding, or varied slightly. According to Webb, analysis of the light coming from distant quasars suggests that a fundamental physical constant may have been increasing slightly over the past six billion years.

The so-called fine structure constant, ε, which measures the strength with which subatomic particles interact with one another and with light, may have been smaller at earlier times in the history of the Universe. Webb says,

"This has major implications for our understanding of physics. If this is correct, it will radically change our view of the Universe. We have to be cautious, but it could be revolutionary. We have seen something in our data - but is it what we think?"

However, the current value of ε and other constants could not have been very different in the past. A small variation in the fine structure constant epsilon (ε) (also called alpha (α)) would imply that carbon atoms could not be stable, and carbon-based life, such as us, could not have arisen. In fact, the six numbers must all lie within an extremely narrow range. Any large deviation would cause havoc. Protons would decay, nuclei would become unstable, DNA could not form, carbon-based life could not happen, and we simply would not be here.

Scientists have since added several other constants to the list of six numbers, but they have not found a law that requires all these constants to have the values they do. In exasperation, these coincidences have driven scientists like Weinberg to invoke the anthropic principle, which essentially states that we observe the Universe the way it is because we exist. If conditions were any different, no one would be here to ponder them. Weinberg admits that he does not like this kind of argument, but we don't know of any other explanation that comes close. Most scientists do not like this principle, and Rocky Kolb, an astrophysicist at Fermi lab, calls the anthropic principle the duct tape of cosmology, which is not going to be permanent.

A related question is, whether there exists only one set of initial conditions that can produce our Universe. According to Hawking, cosmologists are also investigating whether initial configurations favored by the no boundary in space or time, coupled with anthropic arguments, could lead to the evolution of our Universe. If our Universe evolved only from one initial state, then was it a coincidence, or was it chosen very carefully?

If one does not believe in coincidence, then it leads to the idea of intelligent design for the Universe. Recently, such questions and the anthropic principle have given ammunition to a section of the religious community for advancing the intelligent-design movement. They argue that the complexity of nature can only be explained if some higher intelligence was involved in the design.

Of course, different people interpret it differently. Atheists, who do not believe in God or in providential intervention, believe that it was a sheer coincidence. Theists, who believe in God or in continuous providential intervention, use it to indicate the existence of some cosmic power that has been watching over us. Deists believe in God's one-time providential intervention, just at the beginning of the Universe, before He let the Universe evolve. They might use it to indicate the single intervention by God in the beginning to set the Universe on the present course of evolution.

Agnostics, as well as scientists, choose to neither believe nor disbelieve in God or providential intervention. They would like to have more information. They point out that many natural phenomena – lunar or solar eclipse, the changing of the seasons, or lightning – once thought to be direct acts of God, turned out to have simple scientific explanations. Scientists feel that they are stretching their nets and the limits of their intellectual ability as they tackle the complex structure one finds in nature. However, at times, it seems that they are searching for the answers and solutions for seemingly unsolvable problems.

2.10 Multiverse - More than One Universe

Could there be other universes besides ours? Could there be other universes with a different set of six numbers? We have said that our Universe could not exist if six numbers differed significantly from present values. Could there be an infinite Multiverse, which encompasses many disjoint universes – our Universe being one of them - and multiple forms of life different from our form of life?

Could there be more than one Big Bang, which evolved different universes with different sets of physical laws and different sets of constants? Some might have little or no gravity, others stronger gravity. Some might be stable and long-lived; others might be unstable and short-lived. New universes could be triggered within the interior of a black hole when the black holes collapses, creating space and time that is disjoint and never overlapping our space and time.

Rees gives a fascinating description of such speculations in his book, Before the Beginning. He argues that our Universe could be just one of the trillions of universes in the Multiverse, giving the analogy of our solar system. He says that if Earth were the only planet in the Universe, which just happened to be at the right distance from the Sun to be habitable, it would be absurdly impossible. However, when we realize that there are millions of planets, even in our own galaxy, then one planet out of millions that's friendly to life does not seem so surprising.

Inflation theory, quantum mechanics, and string theory also shed a different light on the idea of the Multiverse and on the events before the Big Bang. Linde at Stanford and Velinkin at Tuft University suggest that, while the inflationary period for our Universe ran out of steam early on, it continued in the other universes. According to quantum mechanics, initially there is the 'vacuum' and the uncertainty principle in play, which causes the vacuum to be unstable. It begins to boil, sort of like water, and tiny bubbles begin to form in the vacuum, which expands rapidly.

Each bubble represents a Universe and string theorists call the set of bubbles a 'Multiverse'. Our Universe is one of these bubbles. As these bubbles expand after the Big Bang, different bubbles acquire different values for the cosmological numbers. These numbers decide the fate of different universes in the Multiverse. One version of string theory predicts $\sim 10^{100}$ different universes according to Kachru - a string theorist at Stanford University.

Belief in multiple universes is like belief in religion. Many scientists do not like the idea of multiple universes. There are also questions about the inflation theory and the validity of the string theory. We still do not quite understand the concept of the quantum vacuum. Some might argue that creating bubbles from a vacuum violates the conservation of energy and matter.

The usual answer given is that that no net energy was required to create the Multiverse from a vacuum, since the sum of the so-called positive energy content and the negative energy is still zero, (just like +1 + (-1) = 0). Thus, our Universe was created out of 'nothing'. What an interesting idea! One has to be very careful, though, while interpreting this result.

Einstein once mused over the question of whether God had any choice in creating the Universe. According to some scientists who advocate multiple universes, the answer seems to be 'yes', since many universes are possible. Cosmologists attempt to describe the evolution of the Universe by applying the laws of physics. However, we need to know the initial conditions to predict this evolution. We know that the deterministic laws of general relativity break down near the Big Bang, because it does not incorporate the random component of quantum mechanics such as, Heisenberg's Uncertainty Principle.

The Universe behaves quite similarly to a casino, and as the dice keep rolling, it does not have a single history but has multiple histories - each with its own probability. This allows a large number of possible histories for the Universe. Feynman's idea of multiple histories for a particle, discussed next, when applied to the Universe, also leads to the notion of a Multiverse.

Feynman was awarded the Nobel Prize in 1965 for proposing a different way of thinking about quantum mechanics. He suggested that each elementary particle could travel from one place to another along every possible path through space-time. Thus, a particle has multiple trajectories or histories. Associated with each trajectory are two numbers, one for wave amplitude and the other for phase. The probability of a particle going from one point to another is determined by summing up all the numbers associated with each trajectory.

A macroscopic object follows only one of the infinite possible paths because the rules of assigning numbers to each trajectory insure that all trajectories, but the one actually followed, cancel out in summation. The idea of multiple histories of the Universe (Multiverse) tied to the anthropic principle might lead to the evolution of our Universe.

According to string theory, the mathematics that link up general relativity and quantum mechanics seems to imply that we live in a 10- or 11-dimensional Universe, with up to seven dimensions somehow rolled up into immeasurably small loops. It also implies that our Universe could have developed in a virtually immeasurable number of ways. Hawking believes that the loop dimensions are important, because they may well determine the fundamental characteristics of our cosmos, such as the charge of an electron or the nature of subatomic interactions.

Hawking, however, believes that physicists should focus just on the scenarios that have three large spatial dimensions. It may sound like the anthropic principle. Hawking preferred to use another term - "the selection principle" - because the selection "doesn't depend on intelligent life".

Hawking would like to run time backward to understand the biggest mystery of the Universe. He endorsed the approach favored by the string theorist, Brian Greene. It consists of looking closely at the irregularities in the fingerprints of the early Universe in the background radiation left behind by the Big Bang. Those irregularities, mapped by probes, such as the Planck spacecraft, may reveal the imprint of our own internal space.

Hawking said that the Universe might represent just one 'bubble' in a cosmic froth - perhaps longer-lasting than some other blips. According to him,

"There seems to be a vast landscape of possible 'internal spaces, or multiple universes, and we live in the anthropically allowed region, in which life is possible..

Chapter 3
Questions & Top Ten Mysteries

3.1 Introduction

One fact stands out as we cross various milestones during our journey through science. Scientists, since the dawn of civilization, have been trying to find scientific explanation for the various phenomena observed in nature. Science has indeed made phenomenal progress towards unfolding the mysteries of the Universe and understanding numerous phenomena in nature.

Nevertheless, the Universe remains mysterious in many ways, and scientists are looking for answers to unfold the remaining mysteries. Despite current theories and rapid progress, we are still far from knowing everything about our Universe. In this chapter, we put together a list of questions and the top ten that remain unanswered in science.

3.2 Remaining Questions & Scientific Inquiry

As we learn more about science, we begin to wonder if it has really answered all the questions. Newton explains gravity and the motion of material objects in terms of mass and acceleration. However, do we really know the true nature of matter, mass, and the actual mechanisms causing motion? Maxwell explains electricity and magnetism, including light waves. However, do we really understand the true nature of electrical charges and fields? In thermodynamics, do we know why the disorder or entropy of a closed system must always increase in this Universe? For that matter, despite Einstein's theories, do we know the true nature of time and space?

What is the basis of Einstein's postulate that the speed of light in 'free' space is constant, and that nothing can travel faster than the speed of light? What is this 'free' space, anyway? Does the Standard Model provide all the answers about the atom and fundamental particles? Do we know why energy comes in fixed quanta? Do scientists understand the real physics behind the mathematical equations of quantum mechanics? What really happened before the beginning and during the first 10^{-43} seconds of the Big Bang? Scientists would answer most of these questions with, "DON'T KNOW."

The process of scientific inquiry and discovery essentially remains unchanged. Scientists observe a phenomenon in nature, formulate a hypothesis to explain it in terms of cause and effect, and test the hypothesis with the observed and measured data. If the data supports the hypothesis, and predictions from the hypothesis are confirmed, we raise the hypothesis to the level of 'theory'. Any new theory must embed the time-tested theories as special cases. If a theory meets the test of time and has far-reaching universal implications, we might raise it to the level of a 'law' of nature.

We must remember that this process of inquiry is important. Science would not have accomplished so much, if we did not keep asking new questions. As we move forward in the scientific process of inquiry to answer a question, new questions usually crop up. Scientists then get busy to find answers to the new questions, and the process continues. One scientist contributes, the next one builds upon it, and the process goes on. Science has achieved great success through this process of inquiry. However, for continued progress, we must shine light, not just on the successes, but also on the numerous shortcomings of current theories.

Before proceeding further, we need to pause and think about the nature of inquiry in science. It is true that we must continuously question new hypotheses and theories and raise genuine questions in science. However, while raising such questions, let us be clear about our objective. Our objective should be to discover hypotheses that are more general and theories that lead toward the ultimate truth. In addition, we should never distort scientifically observed data and raise questions just to criticize someone or some hypothesis.

Ideally, we should pose questions, then seek answers without any prejudices or bias, and go boldly wherever our inquiry leads. Unfortunately, history shows that even some very famous scientists could not get rid of their prejudices and preconceived notions, which restricted the scope of their inquiry and delayed scientific progress.

Einstein once said that the most incomprehensible thing about the Universe is that it is comprehensible. Are there any limits to comprehensibility or scientific explanations? Some questions might remain unanswered in any scientific endeavor, due to the very nature of the scientific process of inquiry and the limitations imposed on science. In addition, once we explain a phenomenon in terms of a cause-and-effect, then we have to find the cause that caused this cause.

Every 'why' is answered with 'because...', then another 'why'" crops up, and the endless chain continues. Such a chain of events might even loop or close in on itself. The laws of nature might even be self-contained. The only final question that would remain would be how this self-contained chain, or set of laws, came about. Was any Supreme Intelligence required to design the chain or the laws? If not, then how did they come into existence? If yes, then where did the Supreme Intelligence come from?

Suppose we do reach a situation and discover a unified theory, which answers all the questions about every phenomenon in nature. Then, one would then like to know: why, how, when and where did this law or theory come from? According to Hawking, we would have to know the mind of 'God'. In other words, we can step back, but we might never be able to go back far enough to find the ultimate answer without looking into the mind of God. When asked, on his 60th birthday, about God, Hawking said that God is just a metaphor for the laws of nature.

Let us now continue our journey and raise some fundamental questions that remain unanswered in physical science. Some scientists might claim that many of the so-called unanswered questions have already been answered. The word 'answer' has different meaning for different people. For example, when a child, or a person without any formal scientific background, asks a question about relativity, we usually answer it in simplistic terms with the basic concepts. Our answer might satisfy a child, but a more learned person might find such a simplistic answer inadequate and highly unsatisfactory.

For example, even scientists disagree whether we have really answered questions about the true nature of mass or matter. A scientist would define it either as inertial property - resistance to motion using Newton's laws of motion, or as gravitational property related to weight or force due to gravity, or in terms of matter comprising of particles having certain energy. All these answers are good, but one might still ask, what is the true nature of mass? One could get into a similar dilemma while trying to answer questions about matter or energy. In short, a satisfactory answer depends on the individual's perspective.

We start this phase of our journey by raising questions about Newton's laws of motion and gravitation. We move on to Maxwell and raise questions regarding electricity, magnetism, and optics, etc. We continue with questions about Einstein's theories of relativity. Finally, we raise some questions about quantum mechanics, the Standard Model for the atom, string theory, and conclude by raising some questions about the Big Bang model for the origin of the Universe.

3.3 Did Newton tell the Whole Story?

As stated in the first chapter, the change in position of an object with time is one of the most fundamental characteristics of the Universe. If we reflect for a moment, we realize that it is responsible for so many phenomena of nature. Newton's laws of motion, discussed in the first chapter, are very successful in explaining the motion of material objects at speeds much lower than the speed of light. However, why are Newton's laws true?

Are there more fundamental causes, hidden within the true nature of matter that forces the matter to obey these laws? For example, the first law appears to be a statement of the general principle of causality. It seems to apply to not just material objects, but also to all the phenomena in this Universe, including all the known force fields. It simply states that there can be no effect without a cause and every cause has an effect.

The second law of motion also raises some interesting issues. It confirms that every cause has an effect, and more specifically, it gives a relationship between the cause and the effect in terms of the properties of matter, especially its mass. However, we need to answer some fundamental questions to clearly understand what happens to an object when we apply a force - the causative mechanism that forces matter to follow the first and second law:

    1)   What is the true nature of matter, energy, space, and time?

2) What is the true nature of the mass of the matter?
3) How does an external force bring about a change in position?
4) How does an object sense the causative force field? How fast and how long does it react to it?

Matter is defined as an object that has mass and occupies space, but this definition again avoids a direct answer to the true nature of matter. Inertial mass has been defined as resistance to change, which does not answer the basic question as to why we have this resistance and what actually causes this resistance to be different for different 'mass' objects? Matter, according to current understanding, is composed of molecules and atoms, which are composed of orbiting electrons around nuclei containing protons and neutrons. The Standard Model goes further and considers even more fundamental particles, leptons and quarks - the building blocks of matter. Nevertheless, some questions remain. What is the true nature of the mass of these particles? Why, and through what mechanism, does matter offer resistance to change?

The third law points to a more fundamental law, applicable not just to matter, but also to the force fields. It states that every action causes an equal and opposite reaction to keep the energy balance in the Universe. For example, when we place our hands on an object and push it with a certain force, there is an equal and opposite reaction as the object tries to push us back with the same magnitude. We still do not know the exact mechanism of this action and reaction. In any case, the first and the third laws of motion can be embedded in a more general law, as follows. There can be no effect without cause and every cause has an effect to maintain the energy balance.

Concerning gravity, addressed by Newton or described by Einstein's theory, its fundamental nature is still not well understood. Gravitational theory, addressed in terms of Newton's law or with the rigor of general relativity, is descriptive in nature. The description does not quite reveal the underlying dynamics. The following questions can be raised regarding Newton's and Einstein's gravitational theory:
1) What is the mechanism for the gravitational force to operate? (Is Einstein's space-time curvature concept adequate?)
2) Why does gravity exert only attractive force, unlike some other forces that can be attractive or repulsive?

3) Why does the universal gravitational constant (G) have the value it does, and can we determine its value directly from the theory?

4) Why is acceleration due to gravity independent of the mass of that object?

5) How does the gravity field make its presence felt (mechanism and speed for curving the space fabric)? Recent measurements reported to the American Astronomical Society on Jan 7, 2005 confirmed that the force of gravity travels at the speed of light.

According to the law of gravity, acceleration due to gravity is independent of the mass of an object. In other words, two objects with different masses fall towards the Earth with the same acceleration (g) from a given height (assuming no air resistance). In addition, from the equivalence principle, gravitational mass equals inertial mass, i.e. resistance to motion. Thus, the object with large mass offers more resistance to motion than the one with smaller mass, keeping gravitational acceleration the same for both the objects. However, why does heavier mass offer more resistance?

To answer this question satisfactorily, one must again understand the true nature of mass. We shall address this issue in the next chapter. We also ask the question, 'Does an object accelerate to "g" instantly (i.e., object picks up acceleration in zero time)?' Some of these questions prompted Einstein to develop the gravitational theory, based on the general theory of relativity. Nevertheless, several questions remain unanswered.

3.4 Questions - Heat, Light, Electricity, Magnetism

Heat is essentially a form of radiation energy and an integral part of the electromagnetic spectrum. It is called heat simply because our body reacts in a certain way to it and feels uncomfortable, just as it would to nuclear radiation. Understanding heat energy and the formulation of the laws concerning various related processes have been the crowning achievements of thermodynamics. Now and then, some people claim to come up with examples of thermodynamic law violation, but on further scrutiny, it turns out to be untrue. Nevertheless, some fundamental questions remain unanswered in thermodynamics.

We still do not have a satisfactory answer to the question as to why the entropy or disorder in a closed process must always increase. We know that we can convert one form of energy into another. For example, we generate heat – a band of frequencies in the spectrum – through friction, when some form of energy is converted into heat. Friction at a microscopic level has to be associated with the collisions of molecules, atoms, protons, neutrons, electrons, and even with the fields associated with such particles interacting through the carrier particles.

Nevertheless, we need better understanding of the exact mechanism through which different forms of energy are converted to heat. We can raise the following questions concerning thermodynamics (many scientists might claim to know the answers to these questions, but it would be nice to have answers that are more satisfactory):

1) Is it ever possible to violate the second law of thermodynamics?
2) Why does a closed system always march towards increasing entropy or disorder?
3) What actually happens at the microscopic level when a different form of energy is converted into heat?

We have also noticed, during our journey, that electromagnetic phenomena play a critical role in this Universe. It makes the Universe visible to us. Maxwell provided a beautiful theory to integrate all the electromagnetic phenomena, including light. However, the following fundamental questions regarding electromagnetic phenomena still need better answers:

1) Why do electric and magnetic fields attract and repel, unlike gravity?
2) Why do the permittivity (ε_o) and permeability (μ_o) of free space have the values they do, fixing the speed of light as c (=1/$\sqrt{\mu_o \varepsilon_o}$)?
3) What is the true nature of charge?
4) What is the true nature of the photon, the carrier particle for the electromagnetic wave, and how did it originate?
5) Why and how does the electromagnetic field propagate like a classical wave, exhibiting diffraction and interference

characteristics, but exchange energy like a classical particle as quanta, observing the laws of conservation of energy and momentum?

A partial answer, given for the last question, is that, since the wavelength of the electromagnetic propagating field is extremely small, the propagation of these waves is indistinguishable from the propagation of the photon particles. Therefore, the photon particles behave like the waves, exhibiting diffraction and interference patterns, when the object struck by the propagating field has dimensions comparable to its wavelength. Similar attempts have been made to answer other questions. However, many scientists want to know the answers at a more fundamental level.

3.5 Relativity, Space, Time, Reality and Causality

Some people still raise questions regarding the postulates of these theories. A few questions generally raised are as follows:
1) Why is the speed of light constant in free space, and why at this particular value?
2) Can energy or information propagate at speeds higher than the speed of light in free space? If not, why?
3) Why does the inertial mass equal the gravitational mass?
4) How would non-uniform acceleration affect Einstein's general relativity theory, which was developed for uniform relative acceleration?
5) How is relativity affected at the quantum level? Why are relativity theories inapplicable to the microscopic world?

Several problems also need to be resolved about the true nature of space, time, reality, and causality. We list some of these questions.
1) Could there be higher dimensions to space or even time?
2) What is the true nature of free space?
3) Is 'space' a continuum or a discrete entity?
4) What is 'quantum space'?
5) What is the true nature of time?
6) Could space and time be discrete, like energy and matter?

7) What happens to the concept of space and time at sub-Planck scale?
8) Is the principle of causality ever violated?
9) What is the true nature of quantum entanglement?
10) What is the true nature of reality, independent of the sensors and the processing units that are used to observe and interpret it?

3.6 Is Modern Physics Modern?

Modern physics has successfully explained so many phenomena in our Universe. However, we still need better answers to some of the questions listed below. Again, we are trying to seek better answers than the current well-known answers.

A. Energy Quanta – Why and what are they?

Planck assumed that energy radiates in quanta, in order to resolve the blackbody radiation dilemma. However, it would be nice to go deeper and answer the following questions:

1) Why does electromagnetic energy come in bundles of fixed energy called quanta?
2) What is the physical significance of the Planck constant (h) that relates energy (E) to frequency (f) through the equation $E = h.f.$
3) Can we determine directly the value of the Planck constant, (h) from theory without any measurements?

B. Photoelectric Effect – What is a Photon?

Einstein, taking a cue from the quantum nature of blackbody radiation, came up with the famous photoelectric equation. It won him a Nobel Prize. However, we need better answers to the following questions regarding the photon particles that were introduced:

1) What is the true nature of photons?
2) Why does a photon have energy but no mass?
3) How do photons and electrons exchange energy?

C. Atomic Model – What are Elementary Particles?

The earlier atomic model answered several questions regarding the composition of matter. It zoomed in on matter and discovered protons and neutrons in the nucleus, with electrons revolving around the nucleus in various orbits. In the Standard Model for the atom, we discovered even more fundamental particles - quarks, leptons, etc. The atomic model assigns certain characteristics, charge, mass, etc. to these particles – but we still do not quite understand the true nature of these characteristics. For example, we can raise the following questions:

1) What is the true nature of various particles, and how do they acquire the assigned properties?
2) What is the true nature of the mass or charge for an atomic particle?
3) Can we calculate the mass and charge of an electron or proton, or the mass of a neutron, directly from theory and not from measurements?
4) What is the exact mechanism for energy radiation from an atom when an electron makes a transition from a high-energy state to low-energy stationary state?
5) Why is the frequency of radiation related to the fixed energies (E_i and E_f) of the two stationary orbits?

We shall raise more questions when we visit the Standard Model for the atom.

D. Matter – What is it?

De Broglie proposed the wave nature of all matter, but did not give a good reason or a rationale behind it. We must understand the true nature of the elementary particles in various atoms and the way they influence the composite behavior of matter. Otherwise, we do not have the complete answer to the question, 'Why does all matter have a wave nature?' We have since discovered the critical role of waves in every phenomenon of nature, and one could easily ask the question, 'Is the whole Universe awash with waves, and is matter just a composite of waves?' Finally, what is 'dark matter'?

E. Quantum Mechanics –Physical Explanation

Schrödinger brought in the quantum revolution in science, while vacationing in Switzerland with his girlfriend. He came up with a seemingly unrealistic differential equation, even including an imaginary coefficient, $i (= \sqrt{-1})$. The scientific world is still reeling from it, trying to understand the basis of this equation and to truly come to terms with its far-reaching implications. We still do not have satisfactory answer to the following questions:

1) What is the fundamental principle behind Schrödinger's equation and how can we derive it from the basic principles?
2) Why are quantum mechanics and relativity theory incompatible?
3) How can we develop quantum gravity equations?
4) Is it possible to achieve superconductivity at room temperature?
5) What is the theory behind ceramics and other high-temperature superconductors?

F. Why should we have the Uncertainty Principle?

One of the most troubling implications of quantum mechanics has been Heisenberg's Uncertainty Principle. It introduces an element of uncertainty, and throws a monkey wrench into a neat deterministic model of the Universe. It says that uncertainty is always present when one tries to measure or observe the position and speed, or energy and time, of a particle. However, we still do not have satisfactory answers to the following questions:

1) Why is the uncertainty principle an inherent part of quantum mechanics?
2) Is there uncertainty in the position and speed, or the energy and time, of a particle, because a particle is a vibrating mode of energy in space?
3) What is the true significance of Planck mass, length, and time?
4) What actually happens below the Planck scale, and how is it related to the uncertainty principle?

G. Standard Model – How Standard?

Scientists moved from the earlier atomic model to the so-called Standard Model for the atom, with the help of quantum mechanics and quantum field theory. In the process, they discovered many more particles and even more fundamental particles called leptons and quarks. However, scientists admit that the Standard Model is not the complete answer. We still do not understand several things. For example, we need better answers to the following questions:

1) Why we do have three generations of quarks and leptons?

2) Are quarks and gluons the final fundamental particles of nature?
3) What is their true nature and how did they originate?
4) Why do quarks confine together and what is the lifespan of a proton?
5) How does gravity fit into the Standard Model?
6) How do subatomic particles acquire mass and charge?
7) Why can't the Standard Model predict a particle's mass or its charge?
8) Why can bosons, and not fermions, occupy the same space?
9) How did various carrier particles for the known fields originate?

H. Unified Field & String Theory – Where are we?

We have discovered four force fields and we hope to come up with a theory that unifies these fields. With the exception of gravity, the remaining three fields were unified in a single theory. Attempts to unify gravity with the other forces (electromagnetic, strong, and weak nuclear forces) or to develop a quantum theory of gravity have been unsuccessful. Difficulties arise due to a lack of understanding at the fundamental level. Theoretical scientists have been resorting to ever-increasing levels of mathematical sophistication and abstraction, as is evident in super-gravity and super-string theories, hoping that it would lead to a theory of everything. However, nature keeps throwing curve balls.

1) Have we discovered all the fields?
2) Why is gravity so much weaker than the other three forces?
3) Is string theory 21st century physics, discovered accidentally in the 20th century, or is it just an exercise in mathematics?
4) Where do the strings - the vibrating strands of energy - come from?
5) How will strings tie to yet undiscovered sub-Planck scale physics?
6) Can string theory be the ultimate theory providing all the answers?

3.7 Some Questions about the Universe

A. Single Theory for the Universe

Starting with Planck length ($\sim 10^{-35}$m) and moving on to nucleus (10^{-14}m), an atom (10^{-10}m), DNA (10^{-8}m), cell (10^{-4}m), animal (10m), and stars, galaxies and the edge of the Universe, we cover distances all the way up to 10^{26}m. On time scale, we go from Planck time $\sim 10^{-43}$ sec all the way to ~ 13.7 billion light years ($\sim 4.32 \times 10^{+17}$ sec) – the time that light takes to travel from the observed edge of the Universe. The question that needs to be answered is:

Can we develop a single theory and understand its origin, which would explain every phenomenon in the Universe, irrespective of the space, time or energy scale at which it is observed?

B. Origin of the Universe –Is the Big Bang the answer?

The term 'Big Bang' was coined to ridicule the idea that the Universe evolved from a hot Big Bang or explosion. However, it has turned out to be the most satisfactory model to describe the evolution of the Universe. Several recent observations lend credence to such a model. Nevertheless, numerous questions remain unanswered. Some of these questions can be listed as follows:

1) Did the Universe really start with a single Big Bang?
2) If the Universe originated at time zero, what happened during the 10^{-43} seconds, i.e. below the Planck's time scale?
3) Does the Universe really have an origin and an end in time and space?
4) What is the true cause of redshift and the Hubble constant?
5) Why is space-expansion accelerating, and how would it end?
6) Does the Universe have any boundary conditions? If so, what?
7) Did the Universe evolve in a deterministic manner or from an inherent uncertainty with multiple histories?

C. Stars and Black Holes – Origin and Evolution

The Big Bang model explains the formation of stars, galaxies, and black holes after making certain simplifying assumptions. However, we still do not have a complete understanding of the complexity introduced in the process of evolution. We still have unanswered questions regarding black holes. Some of the questions can be listed as follows:

1) Can we develop detailed mathematical models, which explain the formation of stars, planets, galaxies, etc.?
2) How and when were black holes formed at the center of galaxies?
3) What role do black holes play in the formation of the stars and galaxies, and in the evolution and the fate of the Universe?
4) How can we resolve the black hole information paradox?
5) What would ultimately happen to black holes?

D. Pre- Big Bang & Inflation - What happened?

Scientists realize that the Big Bang model could not explain observations - namely, flatness, horizon, and the absence of monopoles. They had to come up with so-called inflation theory, describing an inflationary phase before the usual Big Bang to resolve these problems. However, details of the inflation theory are still being developed. Several questions, such as those listed below, remain unanswered.

1) What happened before the Big Bang?
2) Did some kind of Universe exist, if any, before the Big Bang?
3) What caused the Big Bang and inflation?
4) Could there be another way to resolve the flatness, horizon and monopole problems, besides the inflation theory?
5) Do any of the two pre-Big Bang models, proposed by string theorists, represent truth, or are they just a mathematical exercise?

E. Dark Matter & Dark Energy

We come across the presence of the mysterious dark energy ($\sim$73%) and the unknown dark matter ($\sim$23%) in the Universe. At present, scientists are trying to understand the nature of dark matter and dark energy. Understanding dark energy might lead scientists to the elusive final unified theory, unifying all known forces that hold the particles in atoms together and the gravity that shapes space.

In fact, the fate of the Universe depends upon dark matter and dark energy, and so does the future of physics. It may also help define a quantum theory of gravity. We might find some clues to the behavior of tiny quanta of gravitational energy in the acceleration of the Universe, and in the implication of Einstein's theory of gravity, allowing some sort of repulsive effect.

1) We might raise the following questions:
2) What is the true nature and origin of dark matter?
3) What is the true nature and origin of dark energy?
4) How do dark matter and dark energy fit into the Standard Model?
5) What are the particles for dark matter and dark energy?
6) How did the amount of dark matter and dark energy change during the evolution of the Universe?

F. Why have just Six Numbers & Multiverse?

Scientists believe that just six numbers made possible the existence of our Universe. Martin Rees postulated this idea in 1999, and we ask the following questions:

1) Why are there only six cosmic numbers with values making the Universe hospitable to known life?
2) These six numbers require about 20 primary constants related to particles and fields, etc.; how was their value set?
3) Are these six numbers absolute constants or could they vary?
4) What about the Planck constant, the speed of light and the gravitational constant? Are they absolute constants?
5) Is the set of six numbers for our Universe the only set, or could there be more sets corresponding to other stable universes?
6) Do multiple universes exist?
7) Could other universes be friendly to some form of life?
8) How can we prove or disprove the existence of the Multiverse or the disjoint parallel universes?

3.8 Top Ten Mysteries of Science

It is obvious from the preceding questions that science has not answered all questions, nor revealed all the mysteries of the Universe. We've put together here a list of the top ten questions that science must address.

1) If there is a cause for every effect, then what caused the Big Bang? Where does this cause-effect chain end? Put succinctly, does the Universe have an origin or an end in time and space?

2) The interaction of the force fields and the resulting energy transformation, following the laws of nature, is responsible for all change and for the existence of this dynamic Universe. What is the origin of these force fields and the associated energy, and which transformation takes precedence?

3) The contents of the Universe are 4% ordinary matter, including radiation, 23% dark matter, and 73% dark energy. What is the exact nature of dark matter and dark energy, and do these percentages vary over time?

4) If every galaxy has a black hole at its center, what is the true nature and role of such black holes in the formation of the stars, galaxies, their evolution, and in the fate of the Universe?

5) If just six numbers are responsible for the existence of our Universe, then how were their values set? Can another Universe exist with a different set of values?

6) How and why do the Planck constant, the speed of light in free space, and the gravitational constant have the values they have? Can we derive these values directly from theory?

7) The Standard Model of particle physics has enjoyed phenomenal success. However, if we have discovered all the force fields, and the Standard Model still does not include gravity, then how do we generalize it and develop a grand unified field theory?

8) If the deterministic relativity theories can explain the Universe and gravity at the macro-scale and quantum mechanics can explain all the phenomena at the micro-scale, then how do we integrate them?

9) If string theory, with its recent discoveries - dualities, application of quantum mechanics and extensions to multi-dimensional 'branes' - is really the physics of the 21st century, then how do we confirm that it describes reality?

10) If our brain is limited in its computational and analytical abilities, by its hard-wired evolutionary programming, and by

the constraints of time, space, and causality, then is it possible for the human mind to discover the ultimate truth about the Universe?

We shall discuss the most likely resolution of these mysteries in the last chapter.

Chapter 4
Frontiers & the Future of Science

4.1 Introduction

During our journey thus far, we have visited the evolution of science, and come across several unanswered questions. By now, we realize that we are far from the final destination. In fact, we do not even know the final destination. We have yet to discover several new theories and universal laws, which would answer all the remaining nagging questions in science. In this phase of our journey through science, we put together the essence of our current understanding. Then, we visit the frontiers of science, and predict its future.

While visiting the frontiers, we'll make some interesting discoveries. For example, we'll discover that we have not yet been able to formulate a satisfactory theory of quantum gravity, which is supposed to bridge the gap between relativity and quantum mechanics. Next, we visit the frontiers that relate to the major ingredients of the Universe and life - namely, change, space, time, energy, and matter.

These major ingredients flow out of the most important process, which is responsible for what we observe in our Universe. This process of change concerns energy in space and time. We develop and state the Universal Principle of Change (UPC), which governs the process of change. First, we illustrate the application of this principle to material objects, and then, to living objects.

As we explore and extend the frontiers of space and time, we raise the possibility of the quantum nature of space and time. We search for the origin of all the energy in the Universe, including dark energy. We discover, to our surprise, that the answers to several seemingly unanswerable questions might emerge from the understanding of empty space, the vacuum, or the void.

We also propose a Singular Configuration Hypothesis and conjectures, which might answer some of the remaining questions in science. Finally, we predict the future of science, and put together a few large pieces of the puzzles in the big picture.

4.2 Known Science in a Nutshell

Let us summarize our understanding of science. Science attempts to answer the question, *'Where am I?'* It essentially observes various phenomena and discovers the laws of nature, which explain and predict the observed behavior. When science discovers a new fact that does not fit with the known theories and laws, scientists modify them to include the new fact. The new theory supersedes the old one, which had some truth in it, as it explained or predicted a phenomenon under certain conditions. Scientific theories are thus continuously updated or modified. Let us now extract the essence of science from the vast knowledge that scientists have accumulated in each field.

A. Classical
1) The Universe has five basic ingredients: energy, matter, time, space, and information.
2) Our Universe is inter-connected, inter-related, and inter-dependent through the interactions of space, time, energy, and matter.
3) Change, affected through the transformation of energy, is the main characteristic of every phenomenon in this dynamic Universe.
4) The Universe obeys the causality principle. An effect cannot precede the cause. There can be no change (effect) in a region of space without energy exchange or transformation (cause). Every cause has an effect, governed by the laws of nature. Every effect becomes a cause and the chain continues.
5) Total energy in the Universe is always conserved. We can transform one form of energy into another, one type of matter into another, and energy into matter or vice-versa.

B. Relativity
1) All physical laws are the same in every inertial reference frame.
2) The speed of light, emitted from a stationary or moving source, in 'empty' or 'free' space in a 'stationary' system is constant. This leads to the consequence that different observers in two frames

moving at constant speed relative to each other have different measures of time and length.

3) Inertial mass, defined by Newton's second law of motion, and the gravitational mass defined by Newton's law of gravity, are equivalent. This leads to the following consequence. If a reference frame is uniformly accelerated, instead of moving at a uniform speed relative to a stationary frame, then we can consider the accelerating frame to be at rest by introducing the presence of a uniform gravitational field.

4) Special relativity theory concerns relative motion with constant speed. General relativity theory considers relative motion with constant acceleration. General relativity defines gravity in terms of the curvature of space fabric. The presence of a material object affects the curvature of space near it, and it is affected by it.

5) Einstein's relativity and gravitational theories form the basis of cosmology and the Big Bang model.

C. Modern Physics

1) Matter is composed of molecules and atoms, and an atom is composed of subatomic particles: protons and neutrons in its nucleus and electrons revolving in orbits around the nucleus.

2) The Standard Model, based on quantum field theory, goes even further. It describes fundamental particles, quarks, as constituents of protons and neutrons. It also explains the three fields: strong, weak, and electromagnetic, in terms of carrier particles.

3) Energy, instead of being continuous, occurs in discrete bundles, called 'quanta'.

4) Subatomic particles have both particle and wavelike characteristics. Their behavior is governed by Schrödinger's wave equation, which can determine it only in terms of the probabilities.

5) According to the uncertainty principle in quantum mechanics, certain pairs of measurements, position and momentum, and energy and time duration, always have an intrinsic uncertainty associated with them

D. Unified Field Theory and String Theory

1) Quantum mechanics and gravitational theories are incompatible.

2) The four known fields are: gravity, electromagnetic, strong, and weak.
3) Attempts to unify gravity with the other three fields have failed.
4) String theory postulates string-like objects as nature's building blocks. It goes farther than the Standard Model, and considers strings of vibrating energy instead of fundamental particles. It promises to yield a unified field theory, but it has not yet delivered.
5) Finally, scientists still do not understand the true nature of the four ingredients of the Universe: energy, matter, space, and time.

E. Big Bang Model
1) The Big Bang model predicts the evolution of the Universe from a singularity.
2) The Big Bang occurred about 13.7 billion years ago and space is still expanding. Physics for the pre-Big Bang is still unknown.
3) Starting with the Big Bang, scientists explain the formation of atomic particles, matter, stars, galaxies, and black holes.
4) Inflation theory and just six numbers explain the structure and the characteristics of our Universe.
5) Scientists know little about dark energy, believed to be responsible for the expansion of the Universe. We also know little about dark matter, that's needed to prevent galaxies from flying apart.

4.3 Frontiers of Science

The field of science is very vibrant. Thousands of scientists, researching day and night, are extending the frontiers of science, and attempting to uncover the secrets of the Universe. Their efforts include theoretical as well as experimental research that's needed to confirm theory. The current research in theoretical and experimental fields will decide the future of science. Each area in science has its own frontier, and the unanswered questions define the current frontiers of science. Before discussing research at the theoretical front, let us briefly examine the frontiers of experimental research.

With powerful telescopes, we have been able to probe space almost to the era of the Big Bang. With powerful microscopes, we have been able to see the detailed chemical structure of a single molecule, including the chemical bonds. It was recently reported in BBC News that a team from IBM Research in Zurich achieved this feat, using an atomic force microscope, or AFM.

The AFM device acts like a tiny tuning fork, with one of the prongs of the fork passing incredibly close to the sample and the other farther away. When the fork is set vibrating, the prong nearest the sample will experience a minuscule shift in the frequency of its vibration, simply because it is getting close to the molecule. Comparing the frequencies of the two prongs gives a measure of just how close the nearer prong is, effectively mapping out the molecule's structure.

Earlier, scientists outlined the physical shape of single carbon nanotubes using similar techniques. Now, their new technique even shows chemical bonds. Such studies on this scale could help in the design of many things on the molecular scale in electronics or even drugs. However, despite remarkable progress in the field of experimental science, we have not been able to measure the distances, time, or mass near the Planck scale. This is where most of the action took place at the time of Big Bang.

Here, we do not visit all these frontiers. Instead, we visit only those frontiers where developments can drastically alter the course of science

A. Quantum Gravity

One such frontier is the frontier of quantum gravity, which is an extremely important challenge to unifying quantum mechanics with Einstein's general relativity theory. Modern physics has two important distinct areas, each describing the Universe on different scales. The first area is quantum mechanics, which ignores gravity and deals with the behavior of the Universe at micro-scale, e.g., atoms, molecules, and fundamental particles.

The second area is general relativity, which deals with large scales, ignoring quantum effects, and talks about gravity bending and warping space-time. Scientists believe that the quantum theory of gravity would combine these two distinct areas of modern physics, and provide an explanation for happenings at the center of a black hole. It might also answer the question, 'What happened before the Big Bang?'

Quantum phenomenon indicates that two particles can be entangled at two different places, and they must adjust to each other at the same instant, but relativity claims that a signal cannot travel faster than the speed of light. General relativity predicts the existence of space-time singularities.

Quantum theory also has infinities showing up in equations, and the uncertainty problem. Does this imply that these theories are incomplete on their own? According to Penrose, the quantum field theory might 'smear' out the singularities of general relativity in some way. Therefore, there is a very real need for a theory that would be consistent, and explain what happens at those tiny length scales at which neither quantum mechanics nor gravity can be ignored.

Much of the difficulty in merging these theories arises because of the radically different assumptions that these theories make. The classical description of other forces involves fields, such as electromagnetic fields, propagating in space-time. Quantum field theory depends on particle fields embedded in the flat space-time of special relativity. However, in general relativity models, gravity, within space-time, changes as the mass moves.

The gravitational force, described by general relativity, is related to the curvature or geometry of space-time itself. In other words, space-time is not an arena in which physical processes take place. It is a dynamical field itself. General relativity is thus not just a theory of the gravitational field, but it is also a theory of space-time itself.

The theory of quantum gravity must address the quantum nature of space and time. However, the quantum nature of space and time is very obscure, and it opens many challenging notions. In the main current version of canonical quantum gravity, it is suggested that quantities, such as area and volume, be quantized. According to Ashtekar, the underlying structure of space-like could be a combinatorial network than a standard continuum manifold. We shall return to the discussion of the quantum nature of space and time, when we visit the frontiers of space and time.

When we combine the two theories, treating gravity as another particle field, we run into the renormalization problem. This problem arises in quantum electrodynamics, when we try to make sense of the infinite results of various calculations to extract finite answers to properly posed physical questions. Renormalization, initially viewed as a provisional procedure, has been accepted as an important tool in several fields. In quantum electrodynamics, the interactions sometimes evaluate to infinite results. However, we one can remove such problems via renormalization.

Gravity particles attract each other. If we add up all the interactions, it results in many infinite values. These values are difficult to cancel out mathematically to yield sensible finite results. Space-time itself is the dynamic field in gravity. The application of Feynman's concept of multiple histories, a sum over histories of the gravitational field in quantum gravity, is really a sum over possible geometry for space-time.

Let us first point out some of the unusual problems associated with the theory of quantum gravity. First, it is difficult to collect any data. This arises from the fact that the natural scale for quantum gravity is the Planck scale. The Planck length scale ($\sim 10^{-35}$m) and the time scale (10^{-43}s) are extremely small compared to the diameters of an atom ($\sim 10^{-10}$m), proton (10^{-15}m), and quark (10^{-18}m). The Planck energy (10^{19}GeV) is also an extreme value. These values suggest that the effects of quantum gravity might be studied directly only in the immediate vicinity of the Big Bang era.

The study of the violation of Lorentz symmetry is also very important to make gravity compatible with quantum physics. As mentioned earlier, Lorentz symmetry has so far withstood the tests of time, but scientists are now questioning whether it is indeed an exact symmetry of nature. Finally, a successful theory of quantum gravity must reproduce the results of quantum field theory and gravity, and provide physical predictions that closely match known observations.

Conceptual problems also arise, not only from disparities in the bases of general relativity and quantum theory, but also from problems about each of these theories. For example, quantum gravity usually includes quantum cosmology. The idea of a 'quantum state of the Universe' presents conceptual problems about quantum theory, such as the meaning of probability, and the interpretation of the quantum state of a closed system—the Universe in its entirety. These examples come from both quantum theory and general relativity. Thus, the difficulties in formulating a theory of quantum gravity are partly due to conceptual problems and not just a lack of data.

The main current approaches to developing a quantum theory of gravity can be listed as follows:

1) String Theory Approach
2) Loop Quantum Gravity (LQG)
3) Non-commutative Geometry
4) Twister Theory
5) Euclidean Quantum Gravity Theory

1) String theory Approach

String theory is a quantum theory, where the fundamental objects are vibrating strings and not point-like particles. String theory can include both the Standard Model and gravity. It predicts the existence of the particle, graviton. However, the theory is background dependent, provides no clear understanding of the physical vacuum, and it has no predictive powers. String theory resolves the problem of a point singularity with infinite density. It does so by setting a lower non-zero size limit (Planck length scale) for a string, and by setting the upper bounds to physical quantities that increase without limit through novel symmetries.

As stated earlier, the string theory for its formulation requires 11 dimensions and super-symmetry, which predicts a host of new particles. Despite intense research activity, string theory is far from achieving its lofty goals.

2) Loop Quantum Gravity (LQG)

The Loop Quantum Gravity approach has been developed by Smolin et al. It is a quantum theory of space-time, which attempts to combine the apparently incompatible theories of quantum mechanics and general relativity. It applies quantum rules to the general theory of relativity. Smolin considers the gravitational field as the fundamental ingredient and treats it quantum mechanically. He avoids the approach that quantizes electromagnetism successfully, but fails for gravity. Instead, he exploits the fact that the theory of gravity is fundamentally nonlinear.

Smolin suggests that we look at space and time in terms of a lattice structure on a tiny scale. The idea is similar to Wheeler's ideas of space-time foam. In other words, if we look at space and time on a very tiny scale, we no longer have three dimensions of space and one of time. The dimensions are messed up in a complicated way. As a theory of quantum gravity, it is the main competitor of string theory. LQG theory is an attempt to formulate a background-independent quantum theory.

LQG theory is based on building spin networks from quantum loops. In LQG, the fabric of space-time is a foamy network of interacting loops ($\sim10^{-35}$ meters in size, Planck scale), mathematically described as 'spin networks'. The loops knot together, forming edges, surfaces, and vertices, somewhat like connected soap bubbles.

It is a way to quantize space-time while keeping the elements of general relativity. A successful attempt would divide a loop into two loops, each with the original size. In LQG, spin networks represent the quantum states of the geometry of the relative space-time. Einstein's theory of general relativity can thus be considered as the classical approximation of the quantized geometry. It is independent of a background gravitational field or metric. It is formulated in four dimensions (3-D space, plus time).

It is not clear whether LQG yields results that match general relativity in the domain of low-energy, macroscopic, and astronomical realm. Although successful for lower dimensions, it has not been shown yet that LQG can reproduce classical gravity in 3-space and one-time dimensions. In the present formulation, it also seems difficult to include electromagnetic, strong, and weak fields. It remains to be seen whether LQG can successfully merge quantum mechanics with general relativity.

3) Non-commutative Geometry

The formulation of the theory of quantum gravity, based on *non-commutative* geometry, relies heavily on operator algebra. In a simplistic manner, it assumes that our space-time coordinates no longer commute, e.g., $x.y - y.x \neq 0$. For example, the phase space of classical mechanics is transformed into non-commutative phase space generated by position and momentum operators.

4) Twister Theory

Twister theory is the mathematical theory which maps the geometric objects of the four-dimensional space-time (Minkowski space) into a four-dimensional complex space with the (2, 2) metric signature. The coordinates in such a space are called 'twisters'. In special relativity, Minkowski's four-dimensional real-vector space is chosen as the model for flat space-time. It takes into account the constant value of the speed of light. Twister theory was once considered a likely candidate for the theory of quantum gravity, but it is not so popular now.

5) Euclidean Quantum Gravity Theory

Stephen Hawking popularized this approach, which assumes that space-time emerges from a quantum average of all possible shapes. It treats time and space in the same manner. The computer models based on this approach represent curved space-time geometries, frequently using tiny triangular building blocks and four-dimensional generalization of triangles. The information emerges from the collective behavior of such blocks, as each building block size approaches zero.

These blocks have no physical significance, and their shape is immaterial. However, computer simulation reveals that the superposition of such blocks is unstable, and it does not lead to the smooth classical Universe on a large scale. In a modified, causal Euclidean Quantum Gravity approach by Ambjorn, Jurkiewicz, and Loll, they assume a built-in distinction between space and time, assigning a built-in arrow of time to triangles, introducing causality. On a small scale, approximating space-time by triangles takes on a fractal shape.

All the above theories have some common features. They look at gravity at a quantum scale. Space-time is discrete and non-commutative. This is simply because, in quantum mechanics, all the physical observable is discrete. Quantum gravity has its own scale, the Planck scale. Quantum theory is supposed to take over when particle energy is comparable to the Planck energy and the existing theories of physics break down. The Planck scale poses a serious puzzle for such small lengths where the separate notion of time and space disappears, and space-time becomes a kind of quantum foam.

Inflation theory might provide a critical piece of the puzzle. We shall revisit this area in the frontiers of science. Quantum gravity must also include a principle akin to the holographic principle since Hawking implies that all the information about a black hole is on the surface of the horizon. Thus, we need to look at the 2-dimensional surface of the horizon to know everything about the black hole.

This is simply because the entropy of a black hole is proportional to the surface area of the horizon and not the volume. Bekinstin also obtained the upper bound on the maximum information (quantized in Planck scale) that can pass through the surface. Although several physicists, such as Smolin and Ashtekar et al. have injected important new ideas into that debate, we are far from any consensus on the theory of quantum gravity.

The Standard Model also has serious problems beyond the energy scale of electroweak fields' unification. Scientists are also looking at theories and models that go beyond the Standard Model, such as grand unified field theories, string theory, and quantum gravity.

B. Universal Principle of Change (UPC)

As we unravel the mysteries of science, we discover some general principles, applicable to all the fields. We summarized these in the classical part (1-5) in the beginning of this chapter. We focus here on a common thread, change, since it is one of the main characteristics of all phenomena in nature – be it material motion, propagating electromagnetic waves, thermodynamics, quantum mechanics, Universe, or life itself. We defined the concept of change in the first chapter. We noticed that change takes place when different fields or matter interacts and one form of energy transforms into another. It is change that makes us aware of the passage and flow of time.

Change takes place in a region of space only when two or more force fields with stored energy interact with each other. These interacting fields initiate the process of energy transformation, which starts the process of change in that region of space. This process brings about change in the state of an object (e.g. position) in that region of space. The state keeps changing while restoring and maintaining the energy balance at every instant, by exchanging, transforming, and storing energy. Recall that one cannot create or destroy energy, but one can only transform it from one form to another.

The process of change continues until the transient variation of the interacting field, force, or energy stops with time, achieving a final equilibrium state. Transient state implies unsteady variation, which, given a chance, might settle into a steady-state variation (a sustained regular pattern) or a constant value. Examples of an equilibrium state are constant or steady state time-variation in space of position, speed, acceleration, temperature, pressure, etc. In fact, the absolute rest position is impossible in this Universe – according to Heisenberg's Uncertainty Principle, which makes change eternal.

Change proceeds along a chosen path, seeking equilibrium. The principle of least action dictates such a path. It states that, out of all the possible paths, one can identify the one path actually taken. It is because a special property of paths, called the action, takes a minimum value along that path. Knowing the description for any system, we can figure out the action to select the same path. We use the slope of the action, piecing it together from path to path, to find the action itself, as we would connect the slopes to find the height of a hill.

One can easily establish a direct connection between Newton's laws and the principle of least action. In fact, we can construct such a description for every complete system that we understand. When we fail to find a corresponding action, it is usually because we have left out part of the important description of the system. One can then go back and include something that was considered unimportant and left out of the description

The cornerstones of the principle of action are symmetry and locality. Symmetry for this principle implies no change in the important circumstances when the path is changed. It corresponds to the conventional notion of symmetry, i.e. a thing looking the same after we subject it to a certain operation. For example, when there is displacement symmetry in some direction, the so-called momentum (defined as the rate of change of the action, per small change in velocity) remains the same at all points along the actual path.

In other words, it conserves the momentum. Similarly, we give the special name of 'energy' to the conserved quantity associated with the symmetry of shifts in time. The understanding of energy conservation came long before we realized that it was indeed just another momentum.

Suppose we assume that the action depends truly on only the characteristics of the system and of the path (space-time trajectory), and not on how we divide time into little pieces. Then, action is composed of little pieces of bits of time multiplied by a Lagrangian, which characterizes the path at that event. Lagrangian, named after the mathematician, Lagrange, who came up with this way of expressing the physical laws, thus depends on the circumstances and the characteristics of the path. It has the units of energy.

It is amazing to note that one can predict the whole dynamics of falling objects in Earth's gravity from only dimensional analysis, symmetry, and locality, as encoded in an action principle. There are remarkably few terms that one can build into an action that depend on the characteristics of the path, and most other terms either violate some symmetry or are redundant. In short, the principle of least action provides a means of formulating classical mechanics that is more flexible and powerful than Newtonian mechanics.

Symmetry and locality of the dynamics are the keys to the choice of an action. Symmetry and locality are often the only fundamental properties we can observe in cosmology. Time symmetry insures conservation of the total energy during the process of change, and space symmetry insures the conservation of momentum along the path. Indeed, these two make the action principle critical in descriptions of nuclear physics, elementary particle physics, and in cosmology. This principle has also proved useful in general relativity theory, quantum field theory, and particle physics. As a result, this principle lies at the core of much of contemporary theoretical physics.

Strictly speaking, the phenomenon of change always continues, since our Universe is interconnected, interdependent, and various force fields are always interacting to some extent in every region of space. Every change (or effect) due to a cause becomes a subsequent cause that causes another change or effect, and so on. Different force fields, no matter how far apart, always continue to interact with each other in some form or another, causing continual change in the state of various entities.

This is precisely the reason why the Universe is dynamic and change in this Universe is perpetual and eternal. Nature brings about changes in weather, seasons, etc. through field interactions in a region of space by exchanging or transforming one form of energy into another. Man can bring about certain changes by the same process.

An interesting example of motion and energy transformation would be the manufactured electric generator. Suppose we start with steam and send this steam into a steam turbine, transforming its thermal energy into mechanical rotational energy. By coupling the steam turbine to an electric generator, the resulting mechanical force rotates the conductors or magnets inside the generator relative to each other.

This relative motion causes the magnetic field near the conductors to vary with time, which induces an electric field that causes the current to flow through the conductors when the circuit is closed. This causal chain, responsible for the change, starts with heat energy. The heat energy transforms into kinetic energy, which in turn transforms into electromagnetic energy. The total energy remains unchanged, after one considers the losses.

Nature does not solve differential equations. Suppose a sudden gush of wind moves a branch of the tree. A scientist would start formulating differential equations based on the laws of nature to predict how the branch would move, given various parameters, wind direction and force, etc. Scientist might be able to predict the motion, solving these equations However, a scientist cannot predict the exact mechanism(s) through which such a change takes place. Nature, on the other hand, would simply follow the Universal Principle of Change.

Let us formally present here the Universal Principle of Change (UPC), which governs the process of change. This principle appears to form the basis of all the laws governing the process of change. In fact, all the laws of nature should follow from this Principle. We could formally state the Universal Principle of Change (UPC), as follows:

Change takes place in a region of space only when two or more force fields interact with each other, disturbing the present state of equilibrium for matter or field(s). The state of matter or field(s) in the affected region continues to change, following the path of least action and maintaining the energy balance through energy exchange and/or transformation.

A stable state of equilibrium is reached only when there is no more transient variation of the interacting force fields or energy with time. The factors affecting the process of change are:

1) *Current equilibrium state*
2) *Characteristics of matter or interacting field(s)*
3) *Environment*

Each of these factors can be deterministic or random. Randomly varying factors obviously cause random changes.

Our dynamic Universe looks interesting only because of the perpetual changes due to the continuous interaction of force fields. This interaction results in energy exchange or transformation during the process of change. The transformed energy continues to restore balance, and ties down or engages energy, initiating the process of change. We'll now illustrate the application of the Universal Principle of Change (UPC) to both material and living objects.

C. Universal Principle (UPC) & Newton Laws

Let us first illustrate the applicability of this principle to motion. Motion is a change in position in space with time. We observe motion everywhere. In the first chapter, we talked about the motion of objects on Earth and in the sky. Whenever force fields in a region of space interact, the energy transformation to kinetic energy might take place and an object in that region can pick up speed to maintain or restore the energy balance.

For example, the gravitational force field's interaction with matter is responsible for the motion of the planets, stars, and galaxies. It ends up placing them in orbits, which results in centripetal acceleration. The resulting inertial force and the kinetic energy counteract or engage gravity and maintain the energy balance.

- Apple Falling from a Tree

Let us zoom in on Newton's law of gravity and explain the underlying basis for the law in terms of force (field) balance. Consider an apple (before Newton sees it falling) with mass (m) hanging at a height (H) from a tree. It keeps hanging as long as the gravitational force field is balanced by the force field. This field is electromagnetic in nature, and it resists the displacement of atoms from the equilibrium position in the stem of an apple. Gravitational force attracts the apple towards the Earth, but it is not yet enough to disrupt the equilibrium and separate the apple from the tree. However, when the apple ripens, the force that resisted the displacement of atoms weakens at the connection, resulting in a force imbalance. This imbalance eventually plucks the apple from the tree.

The apple starts falling towards the ground, following the path of least action, at constant acceleration (g) because of the gravitational force determined by Newton's law of gravitation. It can be written as, $F = m[G.M/(R+H)^2] = mg$, where M and R are the mass and radius of the Earth, respectively. The inertial force (mass x acceleration) balances the gravitational force. The kinetic energy gained because of the speed of the falling apple continually restores and maintains the energy balance.

In the absence of air resistance or any other force, the apple keeps accelerating at g m/sec^2, never achieving a steady speed. It finally hits the ground with a final speed (v). Just before it hits the ground, its original potential energy (mgH), due to the gravitational field (Energy $=F.H = m[GM/(R+H)^2].H = m.g.H$), is transformed into kinetic energy ($0.5mv^2$) and balanced by it. In fact, throughout the fall, the gravitational energy is continuously transformed into kinetic energy and balanced by it, as the speed keeps increasing. From energy balance, it is easy to see that the final speed (v) with which it hits the ground is $v = \sqrt{(2gH)}$.

The final fate of the apple, after it hits the ground, depends on its condition and the contacting surface, when another sequence of momentum and energy exchange takes place. What happens to the kinetic energy ($0.5mv^2$)? It cannot just disappear. It is transformed into, and stored in, the electromagnetic field, due to the displacement of the atoms near the surface of the Earth. This electromagnetic field continues to interact with the atoms underlying the surface of the Earth. The gravitational energy at the ground attempts to displace the apple (or its atoms) towards the Earth's center of gravity, but it is balanced by the electromagnetic energy due to resistance to the displacement of the atoms.

The same process is responsible for establishing an equilibrium between the gravitational force, ($F = m[GM/R^2]$), experienced by the resting apple on the surface with an increase in the electromagnetic field energy. The increase in the electromagnetic field energy, or transformation to this form of energy, is simply due to the displacement of the atoms all the way to the center of the Earth and the resulting stress. In short, the apple finally comes to an equilibrium position and rests on the ground.

Einstein's gravitational theory would explain this interaction between the apple on a tree and the Earth in terms of the curvature of the intervening space because of the gravitational field. This curvature of space eventually forces the apple to fall towards the ground. When the apple comes to rest on the ground, a similar argument can be made. Newton's theory assumes gravity acts at a distance between objects, whereas Einstein's theory explains the action in terms of the curvature of the intervening space. As stated, Newton's theory of gravitation gives excellent results, unless the speed of an object approaches the speed of light.

- Motion of an Object

Let us illustrate the Universal Principle of Change with yet another example concerning Newton's laws of motion. Consider a 2-kg material object lying at rest on a flat horizontal surface, and in equilibrium with the surrounding space. This object is forced to interact with a force field through the sudden application of a 40 N uniform force for 1 second, after which the force is removed. When fields interact and the associated energy is converted into kinetic energy, it pushes the object. How does the object know how to react? The object does not have the ability to think, but it must obey certain physical laws.

According to the principle of change, the force imbalance disturbs the current state of equilibrium. The force field exerted due to the resistance of atoms of the object, electromagnetic in nature, attempts to counteract the applied force field. Some change must occur to restore the force and energy balance. The change occurs according to the Universal Principle of Change and the new equilibrium, depending on the following factors:

a) current equilibrium state, which is the rest position for the object
b) characteristics of matter and applied force – a rigid 2 kg object composed of atoms and subjected to sudden force imbalance of 40 Newtons (N) in a certain direction
c) surrounding environment - contacting flat surface

Fig. 4.1 shows various quantities, as the process of change preserves the energy balance. Suppose the applied force is in a horizontal direction. The object will seek the path of least resistance. If the contacting surface were in a vertical plane, it would not move, but would impart all its energy to the contacting surface, transforming it to heat, provided the contacting surface cannot move. If the contacting surface is horizontal and has enough friction between the object and itself, the imparted energy might not overcome friction, and it gets dissipated as heat. In other words, the object tries to change its position, but it cannot overcome the resistance to the change in position.

The displacement of atoms of the object distorts the electromagnetic field and results in the generation of heat, transforming the imparted energy into heat energy. Hence, the object does not move. However, if the contacting surface happens to be frictionless, then the object's choice is to move on the flat surface horizontally. Motion itself is the result of electromagnetic field interaction between the atoms of the object, although it is explained in a simple way through Newton's laws of motion, in terms of mass and acceleration

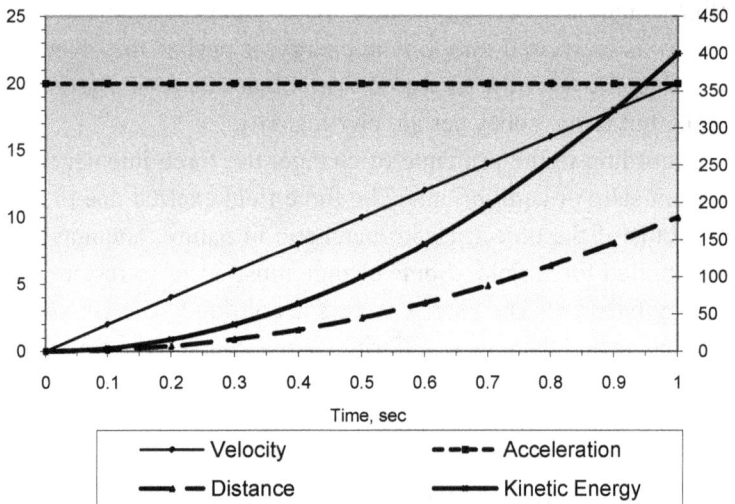

Fig. 4.1 - Motion to maintain energy balance – 40 N force

We know that any motion requires the exchange or transformation of some energy. How much energy is required and what happens to this energy?

As an object moves and covers a distance (x), the energy (E) imparted by the force (F) equals, $E = F.x = (m.a)x$. This energy is converted into the kinetic energy and equals $mv^2/2$ over a distance $x = v^2/2a$. This imparted energy transforms into the acquired kinetic energy. The object travels a distance of $v^2/2a = 10$ m. The acceleration then drops to zero as one removes the applied force after 1 second.

Suppose we wanted to exchange or transform 400 J of energy over a microsecond by applying a constant force of 40×10^6 N over a period of 1 μsec. The corresponding acceleration would be 20×10^6 m/sec^2, the distance traveled would be 10 μm, and the final velocity would be 20 m/s after 1 μsec with the kinetic energy of 400 J, balancing the input energy.

Suppose we imparted this energy as an impulse of strength 400 J, lasting for zero duration (Dirac delta-function). The corresponding impulse force applied would also be infinite for zero duration. The object would pick up instantly at a speed of 20 m/s through infinite acceleration impulse for zero duration to balance the imparted energy impulse of 400 J. After the object has achieved the necessary speed (20 m/s) to restore the balance, it would keep moving at that speed from that point onwards; this becomes its new equilibrium state.

Suppose the object was tiny with a mass of 1.39×10^{-14} kg instead of 2-kg mass? According to energy conservation, it would acquire a velocity $v = (2E/m)^{0.5} = (2 \times 400/1.39 \times 10^{-14})^{0.5} = 2.3983 \times 10^{+8}$ m/s ~ 80% of the speed of light, provided the mass of the object were to remain unaffected. However, we must take into account relativistic effects when the speed of the object approaches the speed of light. For example, if the particle were an electron, the required velocity for balance turns out to be more than the speed of light, according to Newton's classical mechanics, which is not possible, according to Einstein's relativity theory. As the particle accelerates and its speed (v) approaches the speed of light, something must happen.

In fact, an electron approaching the speed of light starts behaving like a photon, and it turns into a photon at the speed of light. According to Einstein's relativity theory, the mass of a neutral particle increases by a factor of $(1-v^2/c^2)^{-1/2}$. It thus achieves a speed that does not exceed the speed of light, which balances the input energy.

If one did not impart the energy at a uniform or constant rate, but through the application of a non-uniform force, the object would accelerate at a non-uniform rate. If one imparts energy continuously by applying a continuous non-uniform force, the object would never achieve a steady speed and the kinetic energy would keep increasing to restore the balance. A horizontally moving object subjected to a continuous force field continues to gain kinetic energy at every instant to maintain the energy balance.

If there is no other means (e.g. friction) available for energy transformation (e.g. to heat), the object has to keep on accelerating, without coming to a steady speed. This is necessary to counteract the force field through inertial force and to maintain or restore the energy balance by continually gaining kinetic energy.

Finally, the random imparting of energy through the application of a randomly varying force would result in random motion. The paper flying randomly due to random wind forces illustrate the same principle of change. If any of the factors involved in the examples illustrating the principle of change were randomly varying instead of being deterministic, the outcome would also be randomly varying with certain probability distributions.

D. Universal Principle of Change & Evolution

- Natural Selection & the Evolution of Life

In my book, Knowing the Unknown – I - *Mysteries of Life..*, we give a detailed discussion about the application of the Universal Principle of Change to the evolution of life. It considers randomly varying factors, and we see that the natural selection principle, the cornerstone of Darwin's theory of evolution, is indeed a special case of the Universal Principle of Change (UPC). One might sum up the principle of natural selection as follows.

Suppose the organisms reproduce, the offspring inherit traits from their progenitor(s), there is a variability of traits, and the environment cannot support all members of a growing population. Then members of the population with less-adaptive traits to survive in the given environment will die out, and members with more-adaptive traits (determined by the environment) will thrive. The result is the evolution of species.

Essentially, the principle of natural selection is an expression of the Universal Principle of Change, where the current state of equilibrium is the certain state of evolution of a set of organisms. The characteristics of the object (living organisms) are the variability of traits and the ability to reproduce offspring and pass on the traits. The environment, besides the usual random factors such as climatic parameter, has the constraints that it cannot support all members of the growing population. Application of the Universal Principle of Change results in the evolution of organisms, continually seeking equilibrium. In fact, one can see this Universal Principle of Change in the field of genetics at almost every level. .

- Natural Selection & Evolution of the Universe

Before we move on, let us point out an interesting twist to Darwin's theory of evolution. Recently, an interesting suggestion has emerged regarding the evolution of the Universe. It states that the structural complexity and constraints coupled with the principle of natural selection might be helpful in explaining the evolution of the Universe. Thus, not all order comes from natural selection alone. It shows that the physical attributes of matter and the structural principles of organization could also produce a great deal of order.

A related interesting idea would be the concept of emergent properties, when putting together complex structures. For example, individual molecules of water, or the atoms of an element, do not exhibit the bulk properties, such as viscosity, etc., which emerge only when nature builds material by putting together all the molecules or atoms. Certain properties emerge out of the complexity and the resulting structure.

The cell is an example of the concept of emergence. A cell is a living independent entity and depends on all sorts of complex interaction within itself. It divides, transforms, and differentiates itself into new types of cells, leading to complex organisms like us. This is called emergence, since it forms something new that is different from its constituent parts. The whole point of emergence is that one does not need something new to form something new.

Smolin thinks that we can create a theory of the whole Universe, which would explain its evolution. It would discover methods by which natural selection might operate on the cosmic scale. The idea advanced by Smolin et al is the 'natural selection' of universes, which states that in some sense the universes that allow complexity and evolution reproduce themselves more efficiently than the other universes. Smolin also believes that the ensemble of universes may evolve inside black holes, not randomly, but by some Darwinian selection, in favor of the potentially complex universes.

Although, physicists and biologists need to work together on this, Brockman says that it is difficult for them to work together. In physics, the theory is more important than the experiments that test the theory. However, in biology, data and the experiment are more important than the theory, because of random variability, since most of the biological parameters in the Universal Principle of Change are randomly varying. Thus, theory comes after collecting the data.

In physics, if the theory appears correct but experimental results differ, then sometimes physicists blame it on experimental error. For example, Eddington measured the bending of starlight by the Sun during a solar eclipse to test Einstein's general relativity theory. When Einstein was asked what he would do if Eddington's measurements failed to support his theory, Einstein's comment was,

"Then I would have felt sorry for the dear Lord. The theory is correct."

4.4 Extending the Frontiers of Space and Time

We visit and extend the frontiers of space and time. We know that energy and matter are the prime stars in this Universe, which perform on the stage of space and time. We can transform matter into energy and vice-versa. In fact, we have chased these two actors down to what we consider the lowest discrete levels, quanta of energy and fundamental particles. We know now that energy occurs in quanta and different elementary particles make up matter. We also associate different carrier particles with different fields.

The three force fields - the strong force, the electromagnetic force, and the weak force - have already been described in terms of quantum theory.

Gravity is the only known force that has not yet been described in terms of energy quanta. Scientists are searching for the carrier particle for the gravitational field. Currently, we do not really know why these field quanta exist, or what exactly matter - or carrier particles - are made of. We also wonder if science would discover more energy fields and particles.

As stated in the previous chapter, science does not have clear answers to many questions. Where did the energy for Big Bang come from? How do different fields emerge from or merge into a single field? How did bosons or carrier particles (e.g. photons) and fermions (e.g. electrons) come into existence? Quantum gravity, a theory yet to be formulated, is supposed to bridge the gap and connect quantum physics and gravity. We know also very little about dark matter and dark energy. Scientists expect a great deal from string theory, but even string theory might not provide the final answer. The origin of strings, or 'branes', itself would require an explanation.

What about the other two ingredients of the Universe besides energy and matter, namely, space and time? Do they just provide an arena where energy and matter perform their act? Could space and time be the prime stars, with energy and matter as the guest stars? At least, could space and time be as important as energy and matter, if not more so?

We know that space affects energy and matter and vice versa. According to Einstein's gravitational theory, the curvature of space (i.e., gravity) decides how matter moves and matter decides how space should curve. Space is an important constituent of the Universe, as it provides the only possible connection among all the observed or predicted phenomena. The other three ingredients of the Universe - energy, matter, and time - exist in space. Therefore, space appears to be the common denominator for all scientific disciplines.

We normally think of space as a container. However, we need to understand and unfold the true origin of space. Only then can we hope to discover the ultimate answers to the remaining questions. The notion of space and time originates at the Planck scale. Therefore, the key to understanding the origin of space and time lies in finding out what happens around the Planck scale. Scientists face problems when they approach the two ends of the space scale, namely, zero and infinite distance. They do not understand the physics below the Planck scale or at the edge of the Universe. We need to find out why space keeps expanding and what the true nature is of vacuum and dark energy.

Some scientists have also proposed a revision of relativity by adding the Planck length scale as the minimum distance, where space-time turns into some kind of quantum foam. However, it poses a serious problem for relativity because one observer sees a particle traveling through ordinary space, continuous and smooth, whereas the other sees it skipping across the quantum foam. Amelino-Camelia in 2000 suggested a revision of the relativity theory and called it doubly special relativity theory.

In this theory, very short wavelengths approaching Planck length start becoming immune to the effect of length contraction, as it causes short-wavelength light to travel at a slightly faster speed than the constant speed of light, c. Smolin, on the other hand, suggests the modification of energy and momentum, when an accelerated particle's energy approaches Planck energy asymptotically; similar to the way an accelerated massive particle approaches the constant speed of light, c.

With this background, let us now visit the frontier of space and time. First, we discuss the history of space, starting with ether, empty space, leading to quantum space, and then review some of the current research activity in quantum space. We also discuss the radical possibility of quantizing both space and time. Next, we propose a 'singular configuration (SC)' hypothesis, and see how it could answer some of the current questions in science.

A. Understanding Space

The concept of space in our mind has always been that of an abstract empty volume. It is hard for us to perceive physical space as concrete, or different from an absolute vacuum. All interactions of matter and energy take place in space and time. We discussed some definitions for space and time in the first chapter. We now ask the questions: What do the terms 'empty' or 'free' space or 'vacuum' space really mean? Is 'vacuum' or 'free' space empty, devoid of everything? Our ultimate objective is to understand the mechanisms of the interactions in space between energy and matter involving motion, gravity, electromagnetic field propagation, etc.

- From Ether-filled to Empty Space

We trace the history of 'ether' first. Ether theory postulates space filled with some sort of fluid. Maxwell believed that electromagnetic waves, like sound waves, need a medium - called 'ether'- for vibrations and propagation. Space ether was assigned properties, such as permittivity, dielectric constant, and permeability, etc. Ether theory had to be abandoned, since the experiments of Michelson and Morley did not confirm its validity. They attempted to measure the effect of the ether drag on the Earth's motion by measuring and comparing the speed of light in different directions. They did not detect any such effect and found the speed of light to be constant.

Einstein also ruled out the existence of ether, and assumed empty space with the associated properties for propagation of electromagnetic waves. Assuming a constant speed of light, his theory of relativity merged the concepts of space and time. According to general relativity, space had a more active role in the formation of the Universe. Space was no longer just an inert container but an integrating part of the Universe, together with matter and energy. However, Einstein still considered space empty. We know from Einstein that the curvature of space is determined by the fields present in space and not by space itself. Similarly, time distortions solely depend on the velocity of moving objects with respect to the constant speed of light.

- Speeding and Accelerating in Space

Speed is experienced relative to another system, but acceleration is absolute and can always be detected as inertial effect. Einstein suggested correctly that Earth's motion be considered relative to other material bodies and not relative to the space surrounding it. An observer, moving with a system relative to another system at uniform speed, experiences no motion unless he observes the second system and not just the space surrounding the first system.

However, if the same observer were accelerating, he would feel the inertial effect, even if he were not observing another system. An example would be spilling a drink in the plane when it suddenly changes its speed.

Alternatively, when we step on the gas paddle to accelerate the car, we are pushed against the backside of the seat.

Mach believed that all motion, including acceleration, is relative. Mach explained this inertial effect of acceleration by suggesting what is now known as Mach's Principle. It states that the inertial effect due to acceleration can be attributed to the influence of all matter in the Universe. Einstein liked this idea, but was disappointed to learn that his general relativity theory did not include Mach's Principle.

In fact, Gödel showed, in 1948, that a solution of Einstein's equation describes a Universe in an absolute state of rotation, not relative to distance matter. The following questions arise: What is acceleration relative to, if it is not relative to the distant matter? Does an object in a system accelerate relative to the space itself, when quantum vacuum pushes against the object, or do we share an absolute reference frame of acceleration with the distant stars? These are intriguing questions concerning acceleration, begging for an answer!

Quantum mechanics also leads us to the phenomenon of 'non-locality,' which implies the following. We cannot characterize an event at a point in space, without reference to the state of the system in the entire Universe. Quantum vacuum provides an explanation for non-locality, since it links local space with global space.

Quantum vacuum connects local physics with distant matter and seems to confirm Mach's Principle, which essentially states that we share an absolute reference frame of acceleration with the distant stars. The space configuration of Higgs' hypothesis also points towards a non-empty space. So does the Ashtekar's team proposal of a space structured in rings, which describes gravitation as a quantum phenomenon.

- Quantum Space & Time Travel

Quantum mechanics also forces us to look at the 'vacuum', or 'ether', again. The 'empty' or 'free' space or 'vacuum' appears to be far from empty. If we took away all the particles and all radiation etc., we should have an empty region of space at a temperature of absolute zero, which is a violation of Heisenberg's Uncertainty Principle. According to this principle, as stated earlier, the more precisely we know the position of a moving particle, the less certain we are about its momentum (mass x velocity) and vice-versa.

A similar uncertainty exists between measurements involving time and energy. The least possible uncertainty is specified in terms of Planck's constant as, $\hbar$ (=h/2 π). This minimum uncertainty is believed to be the intrinsic quantum fuzziness in the very nature of energy and matter.

The uncertainty principle has far-reaching implications. It tells us that the Universe does not know how much energy is associated with a small volume of space for a brief time. The uncertainty principle applies to fields, like the electromagnetic field or the gravitational field. It implies that these fields cannot be exactly zero; otherwise, they would both have a well-defined position at zero, as well as a well-defined speed, which was also zero - a violation of the uncertainty principle. Thus, all the fields must have a certain minimum amount of fluctuations. One could interpret these as quantum fluctuations, like pairs of particles and antiparticles that suddenly appear together, move apart, and then come back together again, and annihilate each other.

In fact, during the early years of quantum mechanics, Dirac theorized, based on the uncertainty principle, that a vacuum was actually filled with particles in negative energy states (Proc. R. Soc. London A, 126, 360, 1930). Pairs of virtual particles, such as an electron and a positron, pop out of nowhere then annihilate each other and give back the energy they borrowed from the vacuum, before the Universe notices them. According to Smolin, quantum fluctuations are statistical fluctuations, due to some unknown universal phenomena. At the same time, quantum fluctuations have certain special properties, which distinguish them from other types of fluctuation phenomena.

A virtual particle in quantum mechanics can never be directly detected. However, its existence has measurable effects. Even during their short lives, virtual electrons are pushed away and positrons are pulled towards the nearby electrons. This polarization of a vacuum has a measurable effect on the properties of electrons. Scientists can measure this effect on electrons, and it matches with quantum theory predictions. In short, space is believed to contain virtual photons and other virtual particles.

Casmir's effect and Lamoreaux's results seem to confirm that the all-pervading vacuum continuously spawns particles and waves that pop in and out of existence. The uncertainty principle limits their time, but they create havoc while they bounce around during the brief lifespan. This active quantum foam extends throughout the Universe, even filling the empty space within the atoms. The Casimir effect is considered to be due to the force produced solely by such activity in the vacuum. Casimir (Phys. Rev. 73, 360, 1948) predicted the presence of such a force and explained why van der Waals forces dropped off unexpectedly at long-range separation between atoms. Lamoreaux (Phys. Rev. Ltrs., 78, 1, 97) verified the Casimir effect using non-conductive plates, as well as conductive plates.

Suppose we place two plates separated by a very short distance. The plates act like mirrors for the virtual particles and antiparticles. The region between the plates behaves like an organ pipe, admitting light waves of only certain resonant frequencies. This results in fewer quantum fluctuations or virtual particles in this region. Since the virtual particles or quantum fluctuations outside the plates can have any frequency, they increasingly dominate the decreasing quantity of virtual particles appearing between the plates.

Because of the reduced number, the virtual particles between the plates do not hit the plates as often as the virtual particles outside. This causes a pressure differential across the plates, and results in a force increasing exponentially with decreasing distance between the plates, $F = K/d^4$. Its frequency dependence is a third power and the force can be altered with dielectrics or resonate with narrow-band mirrors (Phys. Lett. A 225, 1997, 188-194). This force has been measured experimentally. Virtual particles actually exist, and produce real effects.

The Casimir effect also implies that there is lower energy density between the plates than in the outside region. The energy density in the region between the plates is negative, since the energy density of empty space far away from the plates must be zero to prevent space-time from warping and to keep the Universe flat. Thus, quantum mechanics and the uncertainty principle allow negative energy density in some places, provided that it is positive in others. Negative energy density implies that we can warp space-time in the negative direction.

This raises the intriguing possibility of constructing a wormhole, or warping space-time, allowing time travel into the past or future. Such a possibility raises a number of well-known problems, questions, and paradoxes, such as playing havoc with history. If the Casimir effect really implies space-time warping in a micro-region between the plates, it might be that something happens at a macro-scale to prevent time travel. The question of time travel is difficult to resolve indeed!

Casimir also mentioned a three-dimensional volume effect (Phys. XIX, 1956, 846), which has implications of free energy from zero point fields – discussed next. This has been used with the relativistic stress-energy tensor to analyze the quantum electromagnetic field inside any given volume. It was shown that, as the electron density increases due to gravitational compression, there is an energy creation. According to Sokolov, the energy output produced by the Casimir effect during the creation of a neutron star turns out to be sufficient to explain nova and supernova explosions (Sokolov, Phys. Lett. A, 223, 1996, 163-166).

Some scientists believe that the vacuum stress between conducting plates is due to the modification of the electromagnetic zero-point field fluctuations that are discussed next in some detail. One explanation given is that the effect of plates is to reduce the number of wavelengths that fit between the plates relative to the number outside. Thus, the energy density of zero-point fluctuations between the plates is less than the energy density on the outside by a finite amount. Due to this difference in energy, a force arises that pulls the plates together. The New Scientist article (July 1987, "Why Atoms Don't Collapse") points out the importance of zero-point energy (ZPE).

Scientists also believe that space is filled with dark matter, which is at least five times the directly observable matter as revealed by gravitational effects. Space is also believed to contain the electromagnetic zero-point field, the zero-point fields of the weak and strong interactions, and the Dirac's sea of energy and virtual particle pairs. Gravity also affects the energy of quantum space. Scientists also believe in dark energy filling the space. We refer to all of these energies and particles as the quantum vacuum.

- Quantum Fluctuation & Zero-Point Fields (ZPF)

Our discussion thus far points to the fact that space is not a void. It is filled with virtual particles, zero-point field energy, quantum fluctuations, etc. We now discuss zero-point fields and the enormous energy associated with such fields. We shall see that an object does not sense or feel this scalar field as long as it is not accelerating through it. The scalar field also produces small fluctuations because of quantum fluctuations in it, essentially arising from Heisenberg's Uncertainty Principle.

In fact, quantum fluctuation is the temporary change for energy in a point in space, arising from the uncertainty principle. Thus, matter and energy can appear spontaneously out of the vacuum of space, a sort of hiccup in the energy field thought to pervade the cosmos. The energy and time are related $\Delta E . \Delta t \geq \hbar$, which seems to imply that the conservation of energy can be violated, but only for an extremely small duration.

According to the inflation model, quantum fluctuations might have played an important role in the origin of the structure of the Universe. Cosmologists say that a quantum fluctuation gave rise to the Big Bang. Quantum fluctuations can happen anywhere and anytime. Moreover, if our Universe was born out of a quantum fluctuation, then it is possible that other quantum fluctuations could have spawned other universes. We can describe the properties of such a field completely by the power spectrum of the fluctuations.

In inflationary models, the appropriate power spectrum is of a form known as the scale-invariant spectrum. Harrison and Zel'dovich independently derived such a relation for an entirely different purpose in the 1970s. The term `scale-invariant' means that fluctuations in the metric have the same amplitude on all scales (fluctuations in the cosmic microwave background radiation have amplitudes of around 10^{-5}).

We first turn to the zero-point fields. These fields essentially are due to the weak, the strong and the electromagnetic interactions in a vacuum. At present, we do not have a good understanding of the weak and strong interaction zero-point fields. However, Wessen, Hash and others have studied the zero-point field due to electromagnetic waves in the quantum vacuum.

The zero-point field derived by them from the Stochastic Electrodynamics (SED) is somewhat similar to the quantum fluctuations in quantum field theory. However, the zero-point field and the quantum fluctuations are not the same and neither are the mathematical techniques.

Quantum field theory represents the fluctuations as creation and annihilation operators acting on the vacuum. According to Davies, quantum field theory computations yield an energy density of about 10^{110} joules/m^3. However, this cannot be the vacuum energy, since vacuum energy pushes space apart like antigravity, and such an enormous amount would be catastrophic. The work on zero-point field from Stochastic Electrodynamics (SED) is summarized as follows.

An electromagnetic wave is made up of waves of different wavelengths or modes. The field in each different mode swings from one value to another like a pendulum. We can quantize electromagnetic radiation and treat each mode as an equivalent harmonic oscillator. Since each mode of the propagating field is subject to Heisenberg's Uncertainty Principle, each mode of the field must have hf/2 as its average minimum energy.

This is a tiny amount of energy, but the number of modes is infinite in any region of space-time. This energy increases as the square of the frequency. The product of the tiny energy per mode and the huge spatial density of modes yield an infinite theoretical energy density per cubic centimeter.

To make the zero-point field energy finite from the electromagnetic field, they argued that the density of this energy must depend on the frequency, where the zero-point fluctuations cease. As mentioned earlier, space breaks up into a kind of quantum foam at a Planck distance scale ($\lambda \sim 10^{-35}$ m) and in Planck time scale ($\sim 10^{-43}$ sec). They argued that the zero point fluctuations must cease at a corresponding Planck frequency (f $= c/\lambda \sim 10^{43}$ Hz). Even if such were the case, the zero-point energy density would be still 110 orders of magnitude greater than the radiant energy at the center of the Sun.

In their Stochastic Electrodynamics (SED) treatment, the quantum fluctuations of the electric and magnetic fields are treated as random plane waves summed over all possible modes with each mode having this hf/2 energy. Since the flow of radiation is, on average, the same in all directions, there is no net flux of energy or momentum as perceived by an observer in an inertial frame. For example, an object would not be affected by this energy because of its uniform distribution (like an object being bombarded uniformly from all sides). The only effect would perhaps be slight quantum perturbation, somewhat similar to zero-state motion.

Two such effects are attributed to this zero-point field energy. One is the slight perturbation of the lines seen from transition between atomic states known as the Lamb shift. Another is the Casimir effect, a unique attractive quantum force between closely spaced metal plates. Milonni et al., at Los Alamos National Laboratory, showed that the Casimir force is due to radiation pressure from background electromagnetic zero-point energy. This energy has become unbalanced due to the presence of the plates, which results in the plates being pushed together. However, as stated earlier, Lamoreaux verified the Casimir effect using non-conductive plates, as well as conductive plates.

The notion of enormous zero-point field energy, due to fluctuations in quantum space, also causes serious problems in quantum gravity theory. Infinite zero-point energy implies infinite energy density. Like matter, energy density is a source of gravity and infinite energy density would distort space-time fabric and collapse the Universe into a single point – an unreal proposition. Even if the ZPF energy density were finite, its enormous magnitude would generate an enormous space-time curvature, according to general relativity theory. This is, of course, true in the standard interpretation of mass-energy.

The proponents of the zero-point field argue that the ZPF cannot gravitate, because gravitation would involve the interaction of the ZPF with fundamental particles, not with itself. The energy density of the ZPF could then no longer be equated to a source of gravitation. In other words, they get out of this dilemma by assuming that the quantum fluctuations have no gravitational effect. However, the Casimir effect seems to indicate otherwise. Since a force is a source of gravity in general relativity, like matter, we cannot rule out the gravitational effect of this zero-point energy difference.

Most scientists believe that this infinite energy cannot be physically real, and so they subtract infinities in calculations. In fact, Feynman, Schwinger and Tomonaga developed a way of removing or subtracting out infinities. Zero-point or ground state fluctuations, after removing infinities, still caused small effects that could be measured and agreed with experiments. Similar techniques were also used when Yang and Mills extended the Maxwell theory of electromagnetic waves to describe interactions of the strong and weak fields.

Earlier, we also discussed other ways to cancel out these infinities. For example, in the Standard Model, the concept of supersymmetry provides the means to cancel out these infinities in supergravity theories. A consequence of supersymmetry, discussed earlier, is that every particle, fermion or boson has a superpartner with a spin that is ½ less or greater than its own. Furthermore, the ground state energy of a boson (integer spin) is positive, whereas the ground state energy of every fermion (half-number spin) is negative. Since there are an equal number of fermions and bosons, the infinities arising due to ground state energies cancel out.

We have discussed another way to cancel these out in string theory. Another way to take care of infinities would be to go to Einstein's cosmological constant and to assume an infinite negative value for it to cancel out the infinite positive values of the zero-point or ground state energy in space.

A small minority of scientists accepts energy due to quantum fluctuations as real energy, but it is considered small. According to Dolan, if we consider space-time to be discrete, the zero-point energy of an individual point is about twice the Planck energy (equivalent to Planck mass $\sim 10^{-6}$ gram). Then the zero-point energy of any volume of space is also about twice the Planck energy, independent of the volume, if the individual point fluctuations are statistically independent.

The energy density over any observable volume of space is near zero, because it is the average of a very large number of independent random variables. They argue that even if we assume that the energies of the individual point were dependent and the vacuum energy large in any volume of space, it would still be too small to be observed. According to Dolan, this averaging effect is well known in the fields of instrumentation and communications in noise reduction by averaging.

An observer in inertial frame does not experience any effect of the zero-point field. This quantum vacuum, a new version of 'ether' would not exert frictional drag on a particle moving with constant velocity. However, an observer in an accelerating frame, called a Rindler frame, would experience an effect, since acceleration through the quantum vacuum results in the appearance of an electromagnetic Rindler flux.

It would affect the accelerating particles, simply because acceleration is absolute and it can be detected as an inertial effect. For example, as suggested by Davies, an electron circling an atom is jostled by virtual photons from a vacuum, which leads to a slight but measurable shift in its energy. A particle accelerating through the quantum vacuum would find itself surrounded by electromagnetic radiation, like from a hot object.

Could the stretching of space turn some virtual particles into real particles? If true, the expanding Universe could theoretically create particles out of a pure vacuum, as suggested by Parker. A black hole, which is essentially highly warped empty space, when rotating, may fiercely drag and shear ether close to it, causing friction and heat to make it glow faintly. Pandry, at Imperial College, recently showed the same effect by sliding a mirror sideways, parallel to another mirror in a vacuum. This heated up the mirror surface as the kinetic energy of the plates affected the virtual photons between the parallel plates.

Are these effects too weak to have a significant influence in the Universe? Haish and his colleagues claim that the accelerating charged particle in zero-point field (ZPF), due to electromagnetic fields in quantum space, is the true origin of inertia, as it can mimic the effect of mass. As stated earlier, acceleration through the quantum vacuum results in the appearance of an electromagnetic 'Rindler' flux. The fundamental particles (quarks and electrons) in an accelerating object interact with the Rindler flux, generating a drag force proportional to acceleration.

This quantum vacuum inertia hypothesis could thus be the origin of inertial mass (m) in Newton's second law of motion, F=m.a. Thus, inertial mass may not be a physically real property of matter, but resistance of the quantum vacuum to acceleration. This would answer the question of what the acceleration is relative to. According to this notion, we spill a drink in the plane when it suddenly changes its speed because the quantum vacuum pushes against the accelerating atoms.

As far as the rest mass in $E=mc^2$ is concerned, according to Haish and his colleagues, a fundamental particle may be an intrinsically mass-less entity. (Is it string mode, space-time deformation, singularity, or vortex in a perfect, fluid-like medium?) This particle interacts continuously with the quantum vacuum. The particle exhibits Brownian-like motion due to the zero-point fluctuations of the electromagnetic quantum vacuum, named 'zitterbewegung' (quivering motion) by Schrödinger. A tiny bit of the quantum vacuum energy is diverted into the kinetic energy of this zitterbewegung, which could be the origin of $E=mc^2$ for a particle.

The physically real thing would then be only the energy (E) associated with the zitterbewegung of the particle. There is no need for the mysterious conversion of mass into energy and vice-versa. We could think of a particle as a localized concentration of zero-point energy, which gravitates and resists acceleration for the reasons given just now, and no traditional 'mass' is needed. Accordingly, as the number of fundamental particles in a given volume of space increase, so does the energy deficit of the electromagnetic quantum vacuum, since more of it is diverted into zitterbewegung.

This creates an asymmetry in the energy-momentum flow of the zero-point. Such an asymmetry may be perceived as a Rindler flux by a stationary observer. Thus, gravitational mass may be a consequence of mass-less particle's zitterbewegung giving rise to a Rindler flux. In other words, Newtonian gravitational field or a general relativistic curvature of space-time produced by mass may be manifestations of asymmetry of a quantum vacuum energy.

Another unanswered question in quantum mechanics is, 'Why should all matter have wavelike characteristics, and how does a particle acquire such wavelike attributes?' As stated earlier, de Broglie postulated in 1920 that the wavelength (λ) of a moving particle would be $\lambda = h/p$, where h is Planck's constant and p the momentum. He also made a second, less well-known conjecture. If we combine the $E=mc^2$ and the $E=hf$ equations, we arrive at the Compton frequency ($f=mc^2/h$). De Broglie conjectured that the Compton frequency in the case of the electron reflected some kind of fundamental intrinsic oscillation or circulation of charge associated with the electron. In The Quantum Dice, de la Pena and Cetto argue that this presumed oscillation instead can also be interpreted as being externally driven by the zero-point fluctuations of the quantum vacuum.

Pena and Cetto point out a very intriguing result. If the electron really does oscillate at the Compton frequency in its own rest frame, then viewed from a moving frame, a beat frequency becomes superimposed on this oscillation due to a Doppler shift. It turns out that this beat frequency is exactly the de Broglie wavelength of a moving electron. The quantum vacuum inertia hypothesis strongly suggests that the interaction between the quantum vacuum and charged fundamental particles (quarks and electrons) occur at specific frequencies or resonance. Therefore, their conjecture is that the resonance involved in giving the electron inertia is the same as conjectured by de Broglie.

This, in turn, could give a moving electron its apparent wave properties via the Doppler shift effect, as mentioned above. According to Pena and Cetto, it is a very appealing picture suggesting connections not only between electrodynamics and mass, but also between electrodynamics and quantum mechanics. In this picture, the zero-point fluctuations drive the electron to undergo some kind of oscillation at the Compton frequency. It is where the inertia-generating interaction takes place and the de Broglie wavelength originates due to Doppler shifts.

We discussed one idea in terms of zero-point energy due to electromagnetic waves in quantum space, which can explain the nature of mass. Physicists have been trying to understand the true nature of mass using several approaches. According to Wilczek, one of the goals of modern science is to eliminate mass as a primary property of matter. For example, the sum of the masses of the 3 quarks constituting protons or neutrons appears to be only 2 or 3 percent of the inertial mass of those protons or neutrons. Thus, most of the mass is attributed to the energy associated with quark motions and gluon fields. How this energy translates into the property of mass through a Higgs field is not understood.

The explanation of proton and neutron masses in terms of the energies of quark motions and gluon fields still does not provide any insight into the exact nature of inertia. In other words, how does this energy acquire the property of resistance to acceleration known as inertia? In the Standard Model also, one explains it in terms of the balance of the calculated energies with the measured masses. However, balancing energy and mass does not explain inertia. If we eliminate mass as a primary quantity, what happens to the concept of momentum (mv) and kinetic energy ($0.5mv^2$)? Note that the momentum and kinetic energy of material objects in motion are simply mathematical numbers that tell us what to expect when we change velocity. However, the real force is experienced only when that velocity changes. Therefore, if we could explain the occurrence of this force, we can eliminate mass as a primary quantity.

What can we say about the questions concerning gravity and the principle of equivalence of inertial and gravitational mass? General relativity attributes gravitation to space-time curvature. Modern attempts to reconcile quantum physics with relativity treat gravity as an exchange of gravitons in flat space-time, like the exchange of virtual photons in electromagnetism. A non-geometric (i.e. flat space-time) approach to gravity is legitimate in quantum gravity. Similarly, another non-geometric approach assumes that the dielectric properties of space itself might change in the presence of matter.

This is called the polarizable vacuum (PV) approach to gravity. Propagation of light in the presence of matter would deviate from straight lines due to the variable refraction of space itself. The other general relativity effects, such as the slowing of light (as judged by a distant observer) in a gravitational potential, would also occur. However, we infer from the propagation of light that space-time is curved in the first place.

The polarizable vacuum (PV) approach to gravity raises the interesting possibility that general relativity may work, not because of the space-time curvature. It may be due to the point-to-point changes in the dielectric (refractive) properties of space in the presence of matter, which create the illusion of geometrical curvature. A polarizable vacuum type of model does not directly relate gravitation to the zero-point field (ZPF), or to the more general quantum vacuum. However, it does provide a theoretical framework to develop the conjecture of Sakharov, which states that changes in the ZPF create gravitational forces. If correct, gravity would then be due to background zero-point energy pressures - a variation on the Casimir effect. However, this conjecture appears to be at odds with the existence of black holes.

Another objection to the zero-point field (ZPF) hypothesis involves the neutrino. The neutrino is truly a neutral particle - unlike the neutron, which consists of three quarks whose charges cancel - and it cannot have mass originating from electromagnetic interaction. Scientists have recently confirmed that neutrinos have mass. It is the first major finding of the US-based Main Injector Neutrino Oscillation Search (Minos) experiment.

The proponents of ZPF argue that there are two additional zero-point fields, which are associated with the weak and strong interactions. The neutrino is governed by the weak interaction, and it is possible that a similar kind of ZPF-particle interaction creates inertial mass for the neutrino involving the ZPF of the weak interaction.

At present, all this is pure conjecture. No theoretical work has been done on this problem. In any case, according to its proponents, this new approach opens the possibility from which certain areas of the Standard Model and quantum field theory could benefit, such as from a fundamental reinterpretation of mass. The challenge is to see whether quantum field theory can yield an analogous interpretation of inertia and how this would relate to the Higgs field. We have visited several ideas floating around on the frontiers of science, which put a new spin on the currently existing theories.

To sum up, scientists have difficulty visualizing the concept of empty space. Several questions pop up about the true nature of space. Is space truly a fundamental quantity or some special configuration of a deeper quantum entity, whose properties are hard to visualize at this stage? How do we detect the true nature of the quantum vacuum and find out its properties? All of these are good questions! Soon, we shall explore an even more intriguing idea about space and time.

B. Quantum Space & the Origin of Dark Energy

We have talked several times about dark energy (~73% of the total energy in the Universe). As confirmed recently from the redshifts of distant supernovae, the Universe is expanding and accelerating. According to Scherrer, the quickening expansion of the Universe at speeds higher than the speed of light could pull galaxies apart, causing them to drop out of sight. In fact, general relativity allows the possibility of the space between objects stretching at a speed faster than light. We attribute the apparent acceleration of the expansion of the Universe to this dark energy residing in space itself. The dark force, responsible for the expansion of space, is believed to be due to the so-called dark energy in space.

The evidence for an accelerating Universe and the missing energy component also comes from the following argument. The dark energy gives zero spatial curvature to the Universe, which is indicated by mapping fluctuations in the cosmic microwave background radiation. As stated earlier, the CMB anisotropy measurements indicate that the Universe is at $\Omega = 1.0 \pm 0.04$.

In a flat Universe, the matter density and energy density must sum to the critical density. However, matter (baryonic and "dark") only contributes about 1/3rd of the critical density, M = 0.33 ± 0.04 - based upon measurements. These measurements include CMB anisotropy, measurements of bulk flows, and measurements of the baryonic fraction in clusters (- recent WMAP data indicates ordinary and dark matter M = 0.27). Thus, according to the latest measurements, more than two-thirds of the critical density is missing. This missing energy density is attributed to dark energy.

Dark energy has created a serious challenge for scientists, as pointed out while defining energy in the first chapter. Perlmutter said recently that dark energy is something we have no clue as to what is causing it, and it does not fit into current physics. Turner said,

"The dark energy mystery may be answered only by precision astronomical observations and not in the physics lab. One of our goals is to test for dark energy and see if this is preposterous because we are just dead wrong, or find out that we really do live in a preposterous Universe."

According to Wilczek,

"It is not completely fanciful to imagine that this problem will play, in twenty-first century physics, a role analogous to that played by the problem of black-body radiation in twentieth century physics. It might require inventing entirely new ideas, and abandoning old ones we thought to be well-established."

Weinburg made the following comments on the importance of discovering the nature of "dark" energy. Until it is solved, the problem of dark energy will be a roadblock on our path to a comprehensive fundamental physical theory.

"It is difficult for physicists to attack this problem without knowing just what it is that needs to be explained - a cosmological constant or a dark energy that changes with time as the Universe evolves -and for this they must rely on new observations by astronomers."

Some scientists revisit Einstein's theory of general relativity to find the source of missing dark energy. As Turner points out, in Newton's theory, mass is the source of the gravitational field and gravity is always attractive. In general relativity, both energy and pressure source the gravitational field. Thus, sufficiently large negative pressure can lead to repulsive gravity, and Einstein's theory can account for the accelerated expansion.

Repulsive gravity is a completely new feature of general relativity. It leads to a prediction of the accelerating Universe, perhaps, as revolutionary as black holes. If the explanation for the accelerating Universe can be found within general relativity, it will be another important triumph. It is also not clear if Einstein appreciated that his theory predicted the possibility of repulsive gravity.

Some believe that 73% of the mysterious dark energy is due to the cosmological constant. Einstein had introduced this constant in his equation of motion for space-time, since he was not happy with the ever-expanding Universe that was predicted by the equation. Einstein proposed certain conditions in space to keep the Universe eternally balanced at a steady state'. When Hubble discovered that the Universe was not static but expanding, Einstein abandoned his cosmological constant and called it his biggest blunder.

Some scientists believe that dark energy is the vacuum energy, which originates in quantum space. According to this reasoning, space vacuum also must have inertia to counteract and balance the expansion force. Thus, energy in the form of vacuum energy comes from space itself, which pushes the space apart and forces a change in the position of the Universe as it accelerates the expansion of space. The cosmological constant was brought back again by the relativistic quantum theories, in which it arises naturally and dynamically from the quantum oscillations of the virtual particles and antiparticles from nothing.

The quantum field theory made it obligatory to bring back the cosmological constant. Turner, in his paper "Dark Energy and New Cosmology", also discusses the relationship between quantum energy and the cosmological constant11. Since both have the same effect, some scientists equate vacuum energy with the cosmological constant Λ, introduced by Einstein in the general theory of relativity to make sure it described a static Universe.

Although mathematically equivalent, the cosmological constant and the one due to vacuum energy are very different conceptually. Einstein's cosmological constant Λ introduced in his general relativity equation $[G_{\mu\nu} + \Lambda g_{\mu\nu} = 8 \pi G T_{\mu\nu}]$ counterbalances the gravitational attraction to have a static Universe. It is a property of space.

On the other hand, the cosmological constant introduced because of quantum theory arises, perhaps, from the virtual particle-antiparticle pairs. It is the energy density of the vacuum, ρ_{vac}, which goes on the right side of Einstein's equation with other forms of energy, matter and radiation, etc. $[G_{\mu\nu} = 8 \pi G (T_{\mu\nu} + \rho_{vac} g_{\mu\nu})]$.

Vacuum energy is almost the perfect candidate for dark energy. This is because it attributes the properties that dark energy must possess. According to Turner, dark energy emits no light and has a large negative pressure (p_x), which is comparable in magnitude to its energy density (ρ_x), $p_x = -\rho_x$, implying that it is more energy-like than matter-like (matter being characterized by $p<<\rho^{\hat{}}$). It is also approximately homogeneous, and it does not cluster significantly with matter on scales at least as large as clusters of galaxies.

Thus, dark energy is qualitatively very different from dark matter. Unfortunately, the contributions of the well-understood physics to the quantum-vacuum energy add up to $\sim10^{55}$ times the present critical density. String theory also offers a hope to explain the origin of dark energy. Some string theorists call string theory a theory of everything, but it has not yet delivered in this regard.

Now, let us refocus on quantum space to find the source of dark energy. As stated, some scientists believe that dark energy could percolate from the vacuum of space. As stated, laboratory experiments show that seemingly empty space is actually seething with virtual particles that wink in and out of existence. This perpetually bubbling vacuum provides a repulsive 'negative gravity'. However, the problem is that this vacuum-energy, when calculated, is so large that it would have blown apart the Universe very long ago.

According to Wilczek, the natural guesses for the vacuum-energy density, or cosmological constant, vary anywhere from 10^{108} from the quantum-gravity (Planck) scale to 10^{96} from unified gauge symmetry breaking. On the other hand, perhaps, it could be as small as 10^{44} eV^4 if low-energy supersymmetry enforces large cancellations. However, in reality, it is no larger than 10^{-12} eV.

Clearly, we need to resolve this disparity and the most profound gap in our current understanding of the physical world. According to Witten,

"It is vitally important to know if the cosmological constant is truly constant, as inferred from these observations – nonzero vacuum energy or if the observations point to some form of cosmic evolution (sometimes called quintessence). Precise exploration of these exciting questions is likely to have a major impact in physics as well as astronomy."

One way out is to assume that vacuum-energy vastly weakens over time and is not constant as imagined by Einstein. Others talk about the zero-point field energy. Whether any of this is real or not, it is obvious that something is happening in a 'vacuum'. Space seems like a substance itself. If space expands as confirmed by recent experiments, the quantum vacuum should behave somewhat like a viscous fluid and possess inertia to generate inertial force to counteract the expansion force. Furthermore, general relativity suggests that the gravitational field is just a curvature or distortion of the geometry of space-time. Curving or bending space puts a strain on the quantum space, leading to friction when this strain changes with time.

As a source of dark energy, some scientists also suggest the idea called 'quintessence' (for 'fifth essence'). It proposes a repulsive field embedded in space, not unlike a gravitational field or a magnetic field. For example, Steinhardt prefers the idea of quintessence to the cosmological constant. Steinhardt also suggests that this repulsive energy senses the presence of matter and changes its strength or distribution to maintain a balance of densities between matter and energy (~1/3 matter and ~2/3 energy). He thinks that this energy could alter its properties over space and across time. It would not be distributed evenly, and it would not remain constant.

According to this hypothesis, this field was created in the early moments of the Universe, along with the other natural forces. As the Universe expanded and cooled, both gravity and quintessence fields weakened with the Universe's expansion. Finally, gravity weakened more than the quintessence, which took control to push galaxies apart. It now stretches across the Universe like a spider web.

Recently, some scientists have even raised the possibility that the mysterious dark energy might strengthen over time. This could cause violent runaway expansion, ripping galaxies, planets and even atomic nuclei apart. According to Caldwell and his colleague, in a paper submitted to Physical Review, this scenario could play out as soon as 22 billion years. The Milky Way would be destroyed 60 million years before this end, and atoms would be torn to pieces in the final 10^{-19} seconds.

According to Caldwell, there is now a third possibility - the 'big rip'. Instead of the Universe either re-collapsing to a big crunch or expanding forever to a state of infinite dilution, in this scenario, the runaway expansion of the Universe could become increasingly violent. Thus, it would stretch the Universe further apart until the light from the stars cannot reach us. Every observer would see the visible Universe around him or her shrink ever faster, eventually down to a point. Finally, even atomic nuclei will be ripped apart and the Universe will end.

Astronomers want to look through the Supernova/ Acceleration Probe (SNAP) at more distant supernovae. They want to track precisely how the Universe's expansion rate has changed, which might narrow the dark energy alternatives. Other astronomers simply say, 'Why worry?' Dark energy is just a basic feature of the Universe. Trying to explain it is as pointless as trying to explain why Earth was the right distance from the Sun for life to develop. It just turned out that way. If it did not, we would not be here to ask the question.

Experimental efforts are also underway. Turner said that the only laboratory up to the task of studying dark energy is the Universe itself. The NASA WMAP space probe provided some interesting results regarding dark energy in early 2005. However, astronomers are eagerly awaiting the launch of the satellite, SNAP, later this decade.

SNAP is expected to make detailed measurements of thousands of supernovae to determine exactly how fast they are moving away from us. From these measurements, scientists hope to work out how dark energy is changing over time and to find out which fate is in store for the Universe. Depending on the nature of the dark energy responsible, the accelerated expansion may start decelerating and collapse, or keep expanding, or the acceleration may become so huge that the expanding space rips apart even atomic nuclei, ending the Universe.

It is obvious from the preceding discussions that scientists are intrigued and thoroughly confused about the nature and origin of dark energy. Unfortunately, it accounts for ~73% of the total energy in the Universe. If we are to gain a good understanding of our Universe, we must find satisfactory answers to the questions concerning dark energy. At present, we do not understand the true nature of vacuum energy and dark energy, including its magnitude.

C. Quantum Nature of Space and Time

As if scientists are not already perplexed with dark energy and quantum space, we raise another intriguing possibility. The current picture depicts space and time as continuous elements. Is it possible that, if we chased these two actors down to their lowest level, they could also turn out to be discrete packets, like energy and matter? In that case, we would have to drastically revise our picture of the Universe. The primary role and discrete nature of space and time might even be responsible for the discrete nature of energy and particles. If all the four ingredients of the Universe were discrete, difference equations, instead of the differential equations, would be more appropriate to describe the phenomenon in quantum space.

In simple terms, what can we do for space and time? Energy and matter are made up of discrete elements, and it seems rather odd that space and time are the only continuous elements. As a start, we could do what we did for energy and matter. Why not assert that both space and time are not continuous, but occur in quanta, just as energy and the particles do? In other words, space and time are singular configurations and exist like a fixed denomination of currency.

We use the word 'singular configuration' here, in the sense of unique and indivisible configuration. Because the physics at the Planck scale are yet to be understood, we are implying that energy occurs in quanta; the fundamental particles assume discrete singular energy configurations, and so do space and time.

There is some evidence that space and time both do exhibit such singular characteristics and occur in quanta. We know that energy (E) occurs in quanta and equals the force (F) times distance (x), force (F) equals mass (m) times acceleration (a= d^2x/dt^2). However, the mass (m) can be expressed as E/c^2 by the famous Einstein equation, where the velocity of light c is fixed. Putting it together simply suggests that space and time must occur in quanta. In addition, when we zoom in on space and time down to the Planck scale - distance ($\lambda \sim 10^{-35}$ m) and time (t= $\lambda/c \sim 10^{-43}$ sec) - we reach the end of the road. The notion of a separate space and time loses the usual meaning below the Planck scale, since the space-time fabric breaks down.

At the Planck scale, the momentum and energy fluctuations and the gravitational field become too strong, curving space-time and distorting the interval one seeks to measure. Thus, at sufficiently high energy, such as Planck energy, space-time should appear grainy to a particle. That would violate relativity, which assumes that space-time is continuous all the way to zero size scale. In other words, something special happens at Planck scale.

In fact, Einstein himself suggested to Michele Besso in a letter before his death,

"I consider it quite possible that physics cannot be based on the field concept, that is, on continuous structures." He went on to say, "Then nothing remains of my entire castle in the air, including the theory of gravitation, but also nothing of the rest of modern physics."

To reconcile general relativity with quantum mechanics, several physicists have also suspected a short distance cutoff, probably related to the Planck scale, as given in Table 4.1.

Planck Scale		
Mass, m_p	2.1767×10^{-8}	kg
Length, l_p	1.6160×10^{-35}	m
Time, t_p	5.3906×10^{-44}	sec
Frequency, f_p	$1.8551 \times 10^{+43}$	Hz
Energy, E_p	$1.2212 \times 10^{+19}$	Ge.V
Temperature,	$1.4170 \times 10^{+32}$	K
Momentum	6.5256	kg.m/s
Force	$1.2106 \times 10^{+44}$	N

Table 4.1 – Planck Scale

The difficulty in resolving the problem of the infinite gravitational redshift in a black hole, predicted by general relativity, can also be resolved if space-time is assumed to be granular. Space-time would thus be like a material fluid, granular at extremely short distance scale, which would possess a preferred frame of reference manifesting itself on this fine scale.

Could the Planck scale represent the quanta of space and time? Suppose we choose Planck length ($l_p = 1.616 \times 10^{-35}$ m) and Planck time ($t_p = 5.3906 \times 10^{-44}$ sec) as space and time quanta, respectively. Using the physical constants, $\hbar$, c, G we can compute the Planck length l_p, mass m_p, time t_p, and the Planck energy, etc, given in Table 4.1. Note that Heisenberg's Uncertainty Principle relations are $\Delta p . \Delta x \geq \hbar$ and $\Delta E . \Delta t \geq \hbar$. Interestingly enough, the product of momentum and length, or energy and time, equals, i .e., $(m_p c) . l_p = E_p \, t_p = \hbar \ (= h/2 \, \pi)$. Simple computations also show that all the fields have the same energy at the Planck's scale, i.e., $E_p = h \, f_p \ (= h/ t_p) = G \, m_p^2 / l_p = m_p \, c^2 = 1.2212 \times 10^{+19}$ GeV.

We have thus selected the units for length, time, and mass in an arbitrary manner. One could have easily chosen Planck length, Planck time, and Planck mass as units instead of meter, second and kilogram. Planck units of length and time arise naturally when we consider the ultimate limits to measurement. The speed of light would then be unity.

However, these extremely small units are not very practical for our normal perception. The quanta in Table 4.1 are given below in terms of the official units of length (meter) and time (second), as defined by The International Bureau of Weights and Measures.

The meter is the length of the path traveled by light in vacuum during a time interval of 1/1299 792 458 of a second.

The second is the duration of 9 192 631 770 periods of the radiation corresponding to the transition between the two hyperfine levels of the ground state of the Cesium atom.

The kilogram is equal to the mass of the international prototype of kilogram.

If space occurs in 'quanta, what could be between or below the quanta? The question is like asking what is below the quanta of energy or smaller than an electron or quark? In blackbody radiation, Planck postulated that a photon at certain frequency (f) emits only if it has fixed quanta of energy, $E = h.f$, and nothing less. The energy quanta are just like fixed denominations in currency. Similarly, a fundamental particle (a singular energy configuration) constitutes a basic unit of matter.

In fact, at a deeper level, energy quanta and elementary particles must represent configurations of energy, which are stable (observable) solutions of the nonlinear equations describing energy and matter-related dynamics. For example, Schrödinger's equation gives stable, singular configuration solutions only at certain quantum levels. Space and time quanta, similar to energy and matter quanta, constitute basic elements of space and time.

Space is assumed to occur in quanta, since it represents a singular configuration at the Planck scale, just as energy quanta and a fundamental particle represent singular configurations at a certain scale. It does not make sense to ask what is smaller, since that would never be observed, being an unstable solution of a related unknown nonlinear equation. A simple way to visualize space is to imagine cubical bricks of extremely small side length ($l_p = 1.616 \times 10^{-35}$ m) put together to build a wall, which would appear continuous at macro-scale.

Just like space, we can also treat time as a singular configuration at Planck scale and use Planck time ($t_p = 5.3906 \times 10^{-44}$ sec) as the quanta of time. Time does not seem to exist by itself in quantum space. The notion of time arises only when space contains energy, or when a particle (a singular energy configuration), moves with certain delay among space quanta. In other words, time is associated with the particle position or energy movement or change. Since the speed of light ($c = l_p/t_p$) is finite, energy or matter at one point in space cannot appear at another point in space at the same instant, because it cannot transfer instantly.

Thus, space imposes a certain delay on energy as it travels through it. The particle, or the singular energy configuration, reveals itself as it moves or transfers from one discrete space element or space quantum to the next. We identify this delay for the successive transitions in quantum space from one to other space quanta, as the quanta of time.

Stated in another way, the energy in the singular configuration for a particle observed at one position transfers across space quanta. It appears as a singular configuration or particle at another position in space after a certain time delay. In short, it disappears at one position and reappears at another position in space after a certain delay or time. The wave-particle duality confirms this behavior.

This motion or transfer of energy between space quanta cannot take place more rapidly than the speed of light. The elementary unit of measure for time is, therefore, represented by the delay experienced by a photon as it moves over the unit of quantum space - Planck length (l_p = 1.616x10^{-35} m) - at the speed of light. The elementary unit of measure for time is thus the Planck time, t_p = 5.3906x10^{-44} sec in the normal units of time.

Space, time, energy, and material quanta can coexist. Energy and matter can be converted into each other, and material quanta can be looked upon as singular energy configurations at the quantum level. Thus, we could consider particles and energy as different aspects of the same entity, namely, singular energy configuration. Certain amounts of discrete energy transferring between space quanta might materialize into certain singular energy configuration(s) that denotes the particle(s), as observed by a sensor. We could also consider particle to be the outcome of energy interaction with one or more space quanta in quantum space.

Furthermore, various particles, such as leptons, quarks, baryons and mesons could involve one or more space quanta in the interaction of energy. The characteristics of the particle, created through this interaction, would depend on the characteristics of the discrete amount of energy. This interaction process also affects the surrounding space quanta, which could even provide an answer to the baffling question about the origin of dark energy, as a part of the energy imparted to space.

Some scientists suggested the notion of quantized space and time as far back as 1932. The hodon and chronon names were proposed for space and time quanta. The idea did not go far when energy quanta equations were applied without success. The quantum nature of space and time has been revived several times since then without much success. The reason for failure can probably be traced to the lack of understanding of the physics near the Planck scale, where the notion of space and time emerges.

We normally treat space and time as aspects of space-time, represented as a 4-dimensional differentiable manifold. A Lorentzian metric on this manifold represents the metrical structure of space-time. Although gravity in relativity implies space-time itself as a dynamic field, one naturally expects that a theory of quantum gravity will itself adopt the manifold conception of space and time.

Some scientists, such as Butterfield and Isham in the UK, suggest that the 'quantum nature of space and time' need not mean abandoning a manifold conception of space and time at the most fundamental level. However, some research programs in quantum gravity do not accept all of the common treatment, especially as regards the dimensionality and metric structure of space-time - the main difference being their use of some type of quantized metric. According to Butterfield and Isham, the two main current programs even suggest that the manifold conception of space-time is inapplicable on the minuscule length-scales characteristic of quantum gravity.

The fact that our Universe is expanding at a faster rate implies the acceleration of the expansion rate of space itself. The structure of space would thus appear to be such that the space quanta somehow are responsible for generating dark energy. This dark energy pushes space apart and accounts for the acceleration of expansion of the Universe. Furthermore as discussed earlier, slight variations in the material density in space, causing gravitational instability, could account for the web structure of the Universe.

4.5 Singular Configuration Hypothesis (SCH)

Einstein was rather modest when he said that he has no special talents, but is only passionately curious. Einstein had a vivid imagination as borne out by his thought experiments. He said,

"I am enough of an artist to draw freely on my imagination. Imagination is more important than knowledge. Knowledge is limited. Imagination circles the world."

Inspired by Einstein and letting our imagination run wild, let us go out on a limb and discuss some novel ideas. We shall present these ideas under the Singular Configuration Hypothesis (SCH). This hypothesis might help provide answers to some of the remaining questions.

Before formulating the hypothesis, let us review what we do or do not know. We do not know the following about the Universe:

1) Pre-Big Bang scenario and energy for the Big Bang
2) Physics below the Planck scale.
3) Manifestation of various fields and matter from the Big Bang
4) Why relativity theories and quantum mechanics are incompatble
5) Nature and source of dark energy and dark matter

We know the following about our Universe:

1) Space-time fabric breaks down below the Planck scale.
2) Quantum fluctuations create virtual particles and antiparticles.
3) Energy and matter are composed of quanta and particles, and energy and rest mass are equivalent concepts, related through Einstein's famous equation: $E = m.c^2$.
4) The Universe is expanding at an accelerated rate.
5) The Universal Principle of Change governs all phenomena.

Keeping these observations in mind, let us step outside the box and go back to the beginning in search of the answers. Apparently, what happened before the Big Bang and during the early Planck era (before $\sim 10^{-43}$ sec) decided the course of events in the Universe.

As stated, a ten billionth of a second after the Big Bang, the size of our Universe was that of a typical living room, and it had an unbelievably high temperature around billions and billions of degrees ($\sim 10^{20}$K). Particles and antiparticles were created and annihilated, as matter and energy were exchangeable. Somehow, the symmetry was then broken, and for every billion antiparticles there were a billion plus one particles. We must focus on the pre-Big Bang and the early Planck era to find the answers.

Scientists have advanced several scenarios for the pre-bang era. For example, Veneziano in 1991, combining T-duality with the symmetry of time reversal, came up with a pre-Big Bang model. The model paints strange scenarios, like the pre-bang Universe being a mirror image of the post-bang Universe, or proposing a violent transition from acceleration to deceleration. String theory also proposes another model called the 'ekpyrotic' model for the Universe.

This model claims that our Universe is one of many D-branes floating in a higher dimensional space, attracting each other and colliding occasionally. The Big Bang could thus be the impact of the collision of another brane into ours. These are theoretical models lacking any experimental verification.

As far as the physics within the Planck era is concerned, we know very little. We just know that space and time have no distinct meaning within this era. The unknown sub-Planck physics and the associated nonlinear dynamics could lead to the generation of space quanta with Planck length, time quanta with Planck time scale, and a field. Some scientists also suspect that the spontaneous symmetry violations did occur then, which had far-reaching consequences for the evolution of the Universe. The Universal Principle of Change has always played an important role. It is shaping, even now, all the events in this dynamic Universe.

We could put forward these ideas as postulates of the Singular Configuration Hypothesis:

1) *Sub-Planck physics and nonlinear dynamics lead to the generation of space quanta with Planck length (lp), time quanta with Planck time (tp), and the associated field with Planck energy (E_p).*

2) *Spontaneous Lorentz symmetry violations at the Planck scale lead to the generation of various fields, fundamental and carrier particles, while quantum fluctuations create virtual particles.*

3) *Force fields, and various particles, created due to symmetry violation, interact continually, seeking equilibrium according to the Universal Principle of Change (UPC), which leads to the*

formation of nuclei, atoms, stars, galaxies and the various phenomena observed in the Universe.

Why do we call it the Singular Configuration (SC) hypothesis? Normally, a singularity in space marks a point where the curvature of space-time is infinite. It is a region of space-time in which gravitational forces are so strong that Einstein's general relativity theory breaks down, and time and space lose their usual meaning. By a 'singular configuration', we mean a configuration that is unique, indivisible, and stable.

These configurations came into existence mostly during the early Planck era. Space and time are twisted and inseparable in such a configuration. It is almost impossible to create these early conditions for this era in the laboratory in order to understand how these configurations were formed. Quanta of energy, matter, space, and time, discussed in the preceding section, are all examples of singular configurations.

At this stage, this hypothesis remains mere conjecture and is more in the realm of philosophy than science. Nevertheless, it might contain some elements of truth. It could, at least, encourage us to re-interpret and look at the known facts from a different perspective. For example, the hypothesis asks us to look at the so-called observed particles in the Universe as singularities, or material quanta, or more precisely, singular energy configurations (s-e-c), not very different from the predictions of quantum mechanics.

The proposed hypothesis should merely change our perspective about the Universe, just as quantum mechanics did. However, the hypothesis should not negate any of the currently accepted models and theories. It should simply extend and embed them as special cases. Nor should the singular hypothesis refute the current observations regarding various aspects of the Universe.

It might provide a more general framework and help integrate the currently available theories, models and observations. Its development might also include some of the elements of evolving theories, such as the loop quantum gravity, Smolin's theories and the string theory. Of course, the ultimate fate of the hypothesis would be decided by whether it contradicts any of the observations regarding the Universe.

Let us briefly discuss these postulates one-by-one. The first postulate simply points out the fact that the impact of physics and the nonlinear dynamics within the Planck quanta needs to be investigated. At present, scientists are struggling to know the physics below the Planck scale governing the nonlinear dynamics. It also points to the fact that strange things happen to the space-time manifold at Planck scale. It appears grainy as we approach Planck length scale from above, and below the Planck scale, the notion of a separate space and time disappears. Furthermore, time and space can only be generated in Planck quanta.

The second postulate is based on the observation that spontaneous symmetry breaking can lead to a scalar field. It can also have other consequences, discussed in the Standard Model and Symmetry, which are critical for the evolution of the Universe. It is also based on the well-accepted concept of the appearance of virtual particles (particles and antiparticles with positive and negative energy) for a brief period before annihilating each other.

Thus, matter and energy can appear spontaneously out of the vacuum of space. The uncertainty in energy and time are related by $\Delta E.\Delta t \geq \hbar$, which means that the conservation of energy can appear to be violated, but only for a small period. The third postulate simply reemphasizes the application of the Universal Principle of Change to all dynamically evolving phenomena.

In the Singular Configuration (SC) hypothesis, we raise the possibility that a single cosmic field, created within Planck quanta and permeating everywhere, is responsible for the existence of all the energy and matter in the Universe. The existence of such a scalar field was discussed in zero-point fields. This field is spread uniformly everywhere in the Universe, including in our bodies, atoms, and in the measuring devices. Quantum fluctuations or perturbations due to the uncertainty principle in this field are similar to the ripples caused by a disturbance in an ocean, which are superimposed on the ocean propagating in time and space.

The ordinary world of matter and energy that we observe is atop this field, like foam or a ship on the surface of a deep ocean. It raises the datum, and just as it does not matter to a ship how deep the ocean is below it, we do not feel the energy of this field. It is also responsible for the creation of the Higgs boson and Higgs field with a drop in temperature, which we cannot sense directly. One might thus speculate that the physical laws below the Planck scale lead to the generation of a new field. The Casimir effect also lends credence to this speculation, since it is a force produced solely by activity in the vacuum.

Perturbations, or quantum fluctuations, in the scalar field can be quite violent. The larger the energy fluctuation, the smaller is the time over which it occurs, according to the uncertainty principle. This would indicate that perturbations from equilibrium are large, and a nonlinear theory is called far. The interaction of the cosmic field and quantum fluctuations across space quanta creates various singular energy configurations, i.e. both material and carrier particles. The energy, remaining after various energy transformations, or exchange to the matter and other fields, drives the expansion of the Universe.

The new field thus generated within Planck volume could create other space quanta in which quantum fluctuations continue, expanding and stretching the space. Speculating further, the stretching of space somehow might turn some virtual particles into real particles before they can annihilate each other - a possibility suggested by Parker. Thus, the expanding Universe creates energy and particles out of a pure vacuum. Part of the vacuum energy of the new field is converted into radiation and matter particles. The remaining so-called dark energy is responsible for the accelerated expansion of space.

How could the hypothesis include the existing models and answer some fundamental questions? To start with, it could explain the origin of quantum strings in string theory, since fundamental quantum strings originate in the space quanta. We describe a fundamental quantum string – an object with a finite spatial extent, which cannot be described in terms of constituents that are more fundamental.

The resulting different singular-energy-configurations over space quanta appear as different particles, similar to the concept of particles in string theory. In string theory, a string is free to vibrate and different vibrating modes of the string make the string appear as different particles.

According to the SC hypothesis, various observed force fields and particles could be manifestations of the cosmic field interacting with quantum space. A photon is a packet or quanta of energy, which is contained in a singular-wave configuration. It propagates at the speed of light in the cosmic field.

Particles in the Standard Model are manifestations of various singular-energy-configurations (s-e-c). Matter, composed of elementary particles, could be defined as composite of singular-energy-configurations (s-e-c). Mass, volume, charge and spin are the attributes assigned to identify a singular-energy-configuration.

Similarly, the proposed approach considers elementary particles in an atom as manifestation of the energy, which is trapped in a stable, singular-energy configuration. An electron, a singular-energy-configuration (s-e-c), can be visualized as revolving in different orbits defined in the Rutherford's atomic model, or more accurately found with certain probability in various energy levels defined in the Standard Model. Various singular-energy-configurations, with a certain amount of trapped energy, appear to us as particles with attributed notion of certain mass, charge and volume, etc.

The result is the manifestation of various atoms, molecules, and inanimate and animate objects. The second and third postulates could lead to the creation of ~4% ordinary matter, ~ 0.005% radiation, ~ 23% hot dark matter, and ~73% dark energy - accelerating the expansion of the Universe.

We can also modify or convert one singular energy configuration to another (just as we presently convert one particle to another particle, element or isotope) through interactions by providing a suitable energy environment (e.g., as is done in various accelerators). A moving electron radiates an electromagnetic field, and if it were accelerated to the speed of light, it would turn into a photon as all its energy is converted to radiation.

It is also possible to convert a composite of such configurations, so-called ordinary matter, into energy by releasing the energy binding the singular configurations by providing proper trigger mechanism through a suitable energy environment. We can associate a mass (m) with a composite of singular energy-configurations. The energy (E) that can be released from this configuration can be expressed as $E = \Delta m.c^2$, where Δm is the reduction in mass and c is the speed of light.

Of course, the Universal Principle of Change leading to and governing different dynamic phenomena has already been amply discussed. According to the Universal Principle of Change (UPC), each of the particles and their composite matter continually seek equilibrium with the changing environment, due to sudden energy imbalance.

They follow the path of least resistance. The living beings with computing and information processing capabilities follow an optimum path instead of the path of least resistance. The optimality criteria are based on the stored patterns inside the brain computer. Thus, we can paint the big picture of the Universe with a broad brush, using the Singular Configuration Hypothesis.

The SC Hypothesis must embed most of the well-established theories regarding the origin and evolution of the Universe. However, it must go beyond the present theories. Could it be possible that this process is ongoing at the edge of the Universe, creating new Planck quanta, creating new space, time and energy? In such a case, the Universe would keep on evolving and expanding, creating more matter, stars and galaxies, etc.

The SC hypothesis obviously needs analytical development. We must explain exactly what happens within the Planck era. What role does spontaneous symmetry breaking play? How do these events and the quantum fluctuations lead to a cosmic field, entities similar to the vibrating quantum strings, different fields, particles, matter and the Universe that we live in?

Such an analytical treatment is a formidable task! It requires the development of physics for the sub-Planck scale, and the development of new equations for the formation and interaction of different singular energy configurations for different particles. Various properties of different particles - e.g. leptons, quarks, etc. - must follow naturally from these equations. Finally, it would require revisiting models for the composition of the atom from the s-e-c or particles. This is a tall order, indeed!

A. SC Hypothesis & the Unanswered Questions

We now revisit various milestones during our journey through science, and see how various laws, models and theories could be embedded in the general Singular Configuration Hypothesis. We also seek plausible answers to some of the unanswered questions in terms of the SC hypothesis, without any analytical treatment.

- Newton's Laws of Motion

Newton's first law of motion is implied by the last postulate of SC. The Universal Principle of Change (UPC) implies that the matter or the composite singular-energy-configurations (s-e-c) field of a material object is disturbed when it comes under the influence of another force field. This principle implies Newton's first law of motion. Strictly speaking, all singular configurations affect each other to some extent, because of their connectivity through the cosmic field in space. However, this effect may be insignificant for a distant field.

The second law of motion is also implicit in the UPC, as illustrated with an example earlier. The causative mechanism for the motion can also be explained with the SC hypothesis, when we treat matter as composite of singular-energy-configurations (s-e-c). The matter, composed of this s-e-c, interacts with the causal force field configuration, disturbing its field. This disturbance pushes the composite s-e-c field (or matter) like a domino effect. The matter thus appears at different points in space as it moves. The interaction between the two fields must take place at the speed of light. Because of this interaction, the object accelerates and reaches the required speed to restore and preserve the energy balance.

Regarding the inertial mass in Newton's second law of motion, let us restate what Wilczek said in a recent article. He makes an interesting observation that mass, as a primary and irreducible property of matter, is about to be dethroned. According to this article, most of the mass of ordinary matter is the pure energy of moving particles, and a small crucial remainder is attributed to the influence of the Higgs field. The SC hypothesis proposes the origin of mass from a property associated with the singular-energy-configurations and their interactions in the cosmic scalar field, not very different from what was discussed in the zero-point fields.

The third law of motion points to a more fundamental law, which is applicable to matter as well as force fields. It states that every action causes an equal and opposite reaction to keep the energy balance in the Universe. We could treat matter as composite s-e-c field. Then, as an applied force field and s-e-c field interact mutually, both the causal and the composite s-e-c field for the matters are disturbed. The equilibrium is established through force and energy balance, which manifests itself as an equal and opposite reaction.

- Newton's Gravitational Law

As regards Newton's law of gravitation, according to UPC, the energy field associated with one material object acts upon another material object (or its associated field). This disturbs the existing state of equilibrium of each object (field), and the interaction of the fields results in a new equilibrium state. Incidentally, this interpretation of the material object, as composite s-e-c, and its interaction with another force field, also unifies the notion of inertial and gravitational mass. In fact, both the laws of gravitation and Newton's second law of motion describe the interaction of the composite of singular-energy-configurations in matter.

- Electromagnetic Field

Most of the questions raised earlier regarding electromagnetic waves could be satisfactorily resolved if we treat the photon as a unique, stable, singular-energy-configuration. The electromagnetic field, composed of these photons (s-e-c), is generated because of the interaction of a cosmic field with quantum space. A photon's s-e-c interacts with the other photons' s-e-c, exchanging energy, which appear to be pushed in space, while continually seeking equilibrium with the local environment based on the UPC.

This interaction and exchange between the two photons takes place at the speed of light across quantum space in the presence of a cosmic field. This results in electromagnetic waves traveling at the speed of light. The continual exchange of energies between these configurations across quantum space needs further investigation. The s-e-c for photon has unique characteristics, which are different from the s-e-c for other particles. For example, it does not exhibit any 'mass' or resistance to motion, since it exchanges all of its energy with the adjacent photon configuration.

Regarding the notion of electric (or magnetic) charge, the existence of the electric charge density,ρ of particles/ matter in Maxwell's electromagnetic wave equations, is normally characterized as follows. It is given in terms of the net divergence (∇.) or outward flow, of the electrical flux density, D, from a closed space region, which is supposed to include the charge i.e.,∇. D = ρ. Similarly, the non-existence of magnetic charge is characterized as ∇. B = 0, where B is the magnetic flux density.

As regards the true nature of charge, it seems to be the concentrated attribute of the unique s-e-c. Once we can show how the electric field is generated from the cosmic field, Maxwell's first equation can define the existence of charge for a particular s-e-c. In the proposed hypothesis, we are asked to change our perspective, and look at the electric field between particles or s-e-c representing interaction of singular-energy-configurations (particles). As mentioned earlier, it is interesting to note that a unique s-e-c assigned to a particle can be reconfigured into different s-e-c(s), if suitable conditions are created.

- Relativity

Let us now turn our attention to relativity. The SC hypothesis does not contradict any of the postulates or findings of the special or general relativity theories on the macro-scale. However, we assume the space-time manifold to be quantized and not continuous or differentiable on micro-scale. Therefore, one would have to derive the theories of relativity for macro-scale, using a different approach. The SC hypothesis helps us to find the cause for some of the relativity postulates being true. For example, the assumption of the speed of light being constant in space and not exceeded is brought out by the quantum nature of space.

Why is the speed of light constant in free space, and why can't it be exceeded? The main pillar of Einstein's theory of relativity is the fact that the speed of light is constant, independent of the frame of reference, and cannot be exceeded due to some miracle of nature. This can be explained through the Singular Energy hypothesis. The quanta of space and time, being the corresponding Planck scales, l_p and t_p, the energy appears to travel across the space quanta at the speed l_p/t_p, which is precisely the constant speed of light. Since space and time can only be generated in Planck quanta, both can only occur as the same multiples of Planck length and Planck time. Their ratio defines the speed of light in space, and it always remains constant.

This would also be true only if the distance or length was to shorten or shrink and time were to slow by the corresponding multiples of Planck quanta. This is a well-known consequence of Einstein's relativity theory, which is attributed to the fact that the speed of light is constant and independent of the reference frames. We know that the notion of time and space does not exist below the Planck scale. The generation of space and time from the sub-Planck dynamics ensures the speed of light to remain constant in free space.

The SC hypothesis also leads to the equivalence principle, namely, the inertial and gravitational masses being equal. The SC hypothesis confirms that relativity theories are applicable only for describing the Universe at a macroscopic but not microscopic level. Quantum mechanics takes over at the microscopic level, since we must consider quantum effects and the uncertainty principle, due to the existence of ground level or zero-point or vacuum energy inside the space quanta.

- Blackbody Radiation

Let us now visit the blackbody radiation – Ultraviolet Catastrophe, namely, the infinite amount of energy concentrated in very short wavelengths. How did Planck solve that problem? He suggested that the energy emission is not continuous, but can only occur in discrete quanta, like fixed denominations of a currency. A group of atoms cannot emit an arbitrarily small amount of energy at very high frequency. Only atoms (oscillators) with a large amount of energy can emit high frequency radiation. Since the probability of finding groups of atoms with such unusually large energies is low, the spectral curve falls at high frequencies (short wavelengths, $\lambda = c/f$).

A plausible explanation in terms of the SC hypothesis could be given, as follows. A stable singular wave-configuration, a solution of a non-linear equation, can be disturbed by any amount of energy. However, like the typical behavior of nonlinear systems, only one discrete energy level (or stable limit cycle) exists for the next stable s-e-c. The earlier s-e-c can transfer to this stable state, while seeking equilibrium based on minimum energy exchange in the given environment.

In other words, a s-e-c seeks another stable configuration, but finds it only when a sufficient amount of energy is imparted to it. Again, this is perhaps due to the nonlinear nature of the field, that seeks and finds the next stable limit cycle only when enough energy is put into the system. We might even trace the origin of energy quanta to the quantization of space and time.

- Photoelectric Effect

As regards the unanswered questions related to the photoelectric effect, one could also give a plausible explanation for the photoelectric effect in terms of the SC hypothesis. The s-e-c associated with a photon at a certain frequency has certain energy as it propagates and interacts with the electron's s-e-c in an atom of a metal. The result can be freeing and emission of the electron's s-e-c from the metal as the interaction of two different s-e-c takes place and equilibrium is sought in the given environment.

- Atomic Model

Why do we have several discrete energy levels in an atom? This question could also be answered in a manner similar to the emission from a blackbody at different frequencies. Simply stated, because of the nonlinear dynamics, the s-e-c defining electrons that are seeking equilibrium can only achieve certain stable limit cycles. As regards a free electron, it is simply a singular configuration moving in space.

Waves are extended and can be at two places at once, but particles cannot. In fact, it has also been said that particles behave as waves when unobserved (i.e. outside of the necessary conditions of experience), but behave as particles when observed (i.e. appearing under the formal requirements of the phenomena). Some scientists say that the diffraction patterns, typical of wave phenomena, are observed due to different particles striking with different probabilities, just like throwing darts on a dartboard. Such questions become moot in the SC hypothesis, since various particles are considered singular-energy-configurations that interact with each other.

- Quantum Mechanics

The SC hypothesis obviously supports the wave nature of matter, predicted by de Broglie. It also provides support for quantum theory and the quantum nature of energy and matter. It asserts that the energy, instead of being continuous, occurs in discrete bundles, 'quanta'. The SC hypothesis extends the quantum notion to matter in the form of composite of singular energy configurations in space and time for various particles. It resolves the wave-particle duality paradox, and admits probabilities for the occurrence of events.

At present, Schrödinger's equation, weird as it seems with imaginary numbers, is accepted as one of the postulates of quantum theory. It cannot be derived from any basic principles. Suppose one thinks of elementary particles as a spatially-localized wave field, and assigns energy and momentum densities. One way to do so is using the classical 4-tensor energy-momentum theory, which requires a Lagrangian density for the field. Then, according to Bodurov, one could derive the nonlinear Schrodinger's equation, while searching for an equation for the evolution of a complex field that has a Lagrangian density.

Perhaps, the underlying physics behind the SC hypothesis and the new notion of mass might lead to a more fundamental theory, and the Schrödinger equation could be a natural consequence of this theory. The SC hypothesis also lends support to Heisenberg's Uncertainty Principle, which plays a critical role in the quantized space. In fact, the quantum nature of time and space in the first postulates leads directly to the inherent uncertainty, regardless of the measurement process.

- Standard Model

As stated earlier, particles in the Standard Model are mere manifestations of the various singular-energy-configurations (s-e-c). One of the reasons for proposing the singular configuration (SC) hypothesis for the elementary particles is the wave-particle duality paradox encountered with respect to the Standard Model of particle physics. We could consider all particles simply as manifestations of different three-dimensional stable standing wave configurations (s-e-c) as indicated by quantum mechanics. Then, these problems and others, such as particles appearing to be in two places at the same time, can be easily resolved. As stated, waves are extended and can be at two places at once, but particles cannot.

Other questions concerning the mass and charge of atomic particles in the Standard Model of particle physics might be resolved in the SC hypothesis. It does point to the possibility that various common attributes, such as charge, mass of different particles (singular-energy-configurations) might be predictable once we understand the generation of these singular-energy-configurations (s-e-c).

Concerning the Pauli Exclusion Principle, a fermionic particle, or a singular-energy-configuration (s-e-c), can have asymmetric wave function due to its spin characteristics or a quantum number, which prevents it from occupying the same space as another fermion. Similarly, a carrier particle, or configuration, can have a symmetric wave function and can thus occupy the same space as another such particle or configuration.

In short, the SC hypothesis admits various singular energy configurations and wave functions with different quantum attributes for various subatomic particles. The SC hypothesis does not contradict any of the current findings in the Standard Model. It admits the possibility of the creation of various subatomic particles. The analytical tools, nevertheless, remain to be developed for studying the interaction of the cosmic scalar field, which is associated with the vacuum energy. We also need to study interactions within the quantized space for the creation of various fields and various singular configurations (particles).

- Four Fields

We know that energy exists in different forms, and it is the cause of dynamics in the Universe. It is the primary reason for the existence and evolution of matter and life in this Universe. Thus far, four main fields have been postulated. Efforts are ongoing in theoretical physics aimed at arriving at a unified description of all the fundamental forces of nature (gravitational, weak, electromagnetic and strong), in terms of underlying local symmetry groups.

We give a plausible explanation for the existence of these four forces in terms of the SC hypothesis. The existence of four force fields could be explained simply in terms of the s-e-c hypothesis, as follows:

Gravitational - A singular-energy-configuration (s-e-c) 'particle' (or its composite 'matter') with a certain amount of associated energy (mass) that interacts with quantum space and manifests a gravitational field because of the distortion in the field and the space fabric.

Electromagnetic – the creation of the photon s-e-c in quantized space from the cosmic field manifests an electromagnetic (EM) field. The nature of the EM field, photon s-e-c and the charge assigned to electron s-e-c depend on the characteristics of the particular type of singular-energy-configuration (s-e-c).

Strong – The interaction of singular-energy-configurations (s-e-c) associated with subatomic particles manifests the so-called strong field. This interaction is through a field at extremely short distances across the quantum space inside an atomic nucleus.

Weak – It is a singular-energy-configuration (s-e-c), interacting in quantized space through the field under certain boundary conditions, manifests weak field inside or outside the atom nucleus. A singular-energy-configuration (s-e-c) or a 'particle' can transform into other s-e-c(s) or particle(s), depending on the boundary conditions. This is also responsible for splitting an isolated 'neutron's-e-c into two s-e-c or 'particles', electrons and antineutrino.

Thus, all four fields are essentially different manifestations arising from a single field under different boundary conditions and interactions of the singular energy wave configurations (s-e-c) across quantum space. As regards Bell's Theorem and causality, the separated s-e-c fields designated as particles are still in contact through the field, permeating the whole Universe and connected across quantum space. That is why several scientists have suggested the notion of the Universe as an indivisible 'whole' entity, like a hologram.

- String Theory

The SC hypothesis is not only similar to string theory, but it also provides a kind of generic explanation for the existence of strings. In string theory, we describe a fundamental quantum string -- an object with a finite spatial extent. A string cannot be described in terms of constituents that are more fundamental. String theory postulates that nature has string-like objects and not different types of elementary particles and fields. Just as a musical string produces distinct notes when it vibrates, this basic string can vibrate in different modes, and each mode can be viewed as an elementary particle.

The concept of particles as singular energy configurations in the SC hypothesis is quite similar to the concept of particles in string theory. The fundamental quantum strings originate in the space quanta. Just as different vibrating modes of the string appear as different particles, different singular-energy-configurations appear as different particles. It appears that the final string theory would have many of the concepts of SC hypothesis embedded in it, or vice-versa.

- Origin of the Universe

Finally, let us see what SC hypothesis might imply about the origin and evolution of the Universe, the Big Bang model, and the inflation theories. The SC hypothesis supplements the Big Bang model and the inflation theory. The inflation theory assumes a new field, called 'inflaton', akin to the proposed cosmic field in quantum space that caused the accelerated expansion of the Universe immediately after the Big Bang. The SC hypothesis postulates that the cosmic field triggers the Big Bang. It creates more space and energy through new space quanta. It leads to the emergence of four fields, and the partial transformation of energy to matter and radiation, and the current expansion of the Universe driven by dark energy.

The physics for the initial stages of the Big Bang, as stated, remains to be developed. Similar to the Big Bang model, one might suggest the following scenario for the evolution of the Universe according to the SC hypothesis. The Universe starts with the initial energy – perhaps the vacuum energy borrowed from the space quanta. At this point, space or time does not exist outside the single space quantum. Due to vacuum pressure, more space quanta are created adjacent to the existing space quanta, stretching the space as the rapid expansion process continues.

Each quantum of space has the following parameters associated with it. Planck scale length $l_p = 1.616 \times 10^{-35}$m, time $t_p = 5.3906 \times 10^{-44}$ sec, with energy $E_p = 1.2212 \times 10^{+19}$ GeV, and energy density $\rho_p = 4.64 \times 10^{+113}$ Joules/m^3. This expansion process proceeds at a rapid pace. It is somewhat like building a mosaic floor with tiles of Planck scale dimensions.

The SC hypothesis has some elements of inflation theory as the space expands at an accelerated rate and more space quanta are created. During the first $\sim 10^{-36}$ seconds, something remarkable happens. A new kind of 'cosmic repulsion' comes into play because of the huge (vacuum) energy density, which exerts huge pressure, overwhelming gravitational force, and accelerates the expansion of space at an exponential rate. It inflates the embryo Universe, homogenizes it and establishes a very fine-tuned balance between gravitational and vacuum kinetic energy. The inflation ends when the vacuum decays into a more ordinary state.

According to inflation theory, the transition releases heat as water releases latent heat on freezing. The heat survives; it is cooled and diluted as $2.728°K$ background radiation. The arguments of the inflation theory, proposed by Guth in his book, The Inflationary Universe, are still somewhat speculative, as they depend on unknown physics at extremely high energies. However, as stated earlier, they do solve the flatness and horizon problem. As we see, inflation theory fits in with the SC hypothesis, which proposes the exponential expansion of space, as space quanta are generated.

According to the SC hypothesis, the standard evolution of the Universe starts at the end of the inflationary era from a small hot dense quantum state. Some virtual particles turn into real particles or various singular-energy-configurations (s-e-c). Thus, the expansion and stretching of space creates particles out of pure vacuum, as suggested by Parker. The most likely reason for this transformation is the interaction of vacuum energy between the adjacent space quanta. Part of the vacuum energy of the quantum space thus transforms into radiation energy, i.e. a soup of photons, gluons and other elementary particles.

The vacuum energy is reduced through energy transformation, cancellation as the Universe expands, and the temperature starts to drop. The Universe then evolves as described by the Big Bang theory. When the expanding Universe grows to 1/1000th of its current size, about 380,000 years after the Big Bang, and the temperature drops enough, protons can capture the slow-moving electrons and become neutral atoms, forming matter. Photons, from dense as well as rarified regions, can now travel without being scattered by collisions with charge particles. According to Hu and Martin, the patterns of hot and cold spots, induced by sound waves, freeze into the Cosmic Microwave Background (CMB).

The exchange between matter and radiation becomes less efficient, and photons behave as thermal, blackbody radiation. This is measured even today as background radiation. After having cooled off for many billions of years, the temperature of this radiation is only a few degrees above absolute zero. As the temperature drops and the average speed of electrons are reduced, electrons are captured by protons and atoms start to form.

The first neutral atoms form from the nuclei of hydrogen, helium, and lithium. Eventually, matter energy density - mostly in nuclei - dominates radiation energy density, forming stars, galaxies, heavier elements in stars, and planets. Thus, matter freed from radiation pressure clumps together under the influence of the gravitational field, forming stars and galaxies.

B. SC Hypothesis & Energy of the Universe
Let us discuss the current energy balance given in Table 4.2.

Material	Typical Particles (s-e-c)	Particle Energy (Ge.V)	Number of Particles	Energy Contribution (Ge.V)
Ordinary Matter & Radiation	Electrons Protons Photons	5.11×10^{-04} 9.38×10^{-01} $\sim 10^{-13}$	10^{78} 10^{87}	$\sim 0.4 \times 10^{79}$ (4%)
Hot and Cold "dark" Matter	Neutrinos ?	$<10^{-9}$ $\sim 10^{2}$	$\sim 10^{87}$ $\sim 10^{77}$	$\sim 2.3 \times 10^{79}$ (23%)
"Dark" energy	"Scalar"?	$\sim 10^{-40}$	$\sim 10^{118}$	$\sim 7.3 \times 10^{79}$ (73%)
TOTAL				$\sim 10^{80}$ (100%)

Table 4.2 – Current Universe energy balance

For illustration purposes, we use the new WMAP estimates for the composition of the current Universe, instead of the numbers in a table in an article by Cline (March 2005 issue of the Scientific American).

Scientists had to come up with lot of additional energy (~73%) - they named it "dark" energy - to account for the rapid expansion of the universe. They also had to come up with additional matter (~23%) - they called it "dark" matter otherwise the gravitational pull would not be enough to hold the galaxies in place. The total energy including ordinary and "dark" matter, radiation, and "dark" energy, required to balance the energy required for expansion and gravitational pull is estimated in Table 4.2. Again, it is interesting to point out that the net sum of energy in the universe is zero, if we denote the energy in the Table as positive, and the needed gravitational energy and the energy required for expansion as negative.

The average energy density of the Universe, according to the total energy from the Table 4.2, would be ~ 46 GeV/m^3 or 7.35×10^{-9} J/m^3. This small energy density is like having a single proton in every cubic feet of space. On the other hand, the Planck energy in the space quanta is 1.22×10^{19} GeV, which yields an energy density of $\sim 2.89 \times 10^{+123}$ GeV/m^3 or $4.64 \times 10^{+113}$ J/m^3. That is a huge gap, i.e., ρ_{vac} (theory) $= 10^{+120} \rho_{vac}$ (observed), which is embarrassing for the proponents of the vacuum energy. They are desperate to find some plausible explanation for this huge discrepancy.

One attempt to resolve this discrepancy, given by some scientists, is as follows. They believe that the vacuum energy does not remain constant and decreases exponentially as space is stretched. Similar to inflation theory, the space volume stretches exponentially, exp $(2/\varepsilon)$, where $\varepsilon = 1/137$. In this context, it is interesting to note that the natural log of 10^{+120} equals 2×137, which by a strange coincidence, happens to be equal to $2/\varepsilon$, where ε is the fine structure constant. Thus, ρ_{vac} (observed) $=$ exp $(-2/\varepsilon)\, \rho_{vac}$ (theory).

Harvey, in a 1999 paper in *Physical Review*, suggested that the non-perturbative effects in non-supersymmetric string theories might lead to such an answer. It is important to note that the above explanations are based on crude data and far-fetched assumptions. This might turn out to be mere coincidence. On the other hand, it could imply that vacuum energy is indeed responsible for all the energy in this Universe, as it is transformed into other forms of energy. Even if it is true, the SC hypothesis at this stage cannot provide the answer to the question concerning the specific distribution of the vacuum energy into various forms.

According to the SC hypothesis, the following scenario could be visualized, similar to the inflation theory, to reconcile the estimated observed energy density with the theoretical energy density. We assume that the space volume stretches exponentially, $\exp(2/\varepsilon)$ where, $\varepsilon \sim 1/137$. The newly created Planck space quanta are also stretched. The volume of such stretched Planck quanta is $\sim 2.65 \times 10^{+17} m^3$. Each space quanta contributes $\sim 1.22 \times 10^{19}$ GeV, which implies an energy density of 46 GeV matching the observed value. The number of such stretched space quanta in the Universe would be $\sim 10^{61}$. Since each space quanta contributes $\sim 10^{19}$ GeV, the total energy contribution due to the energy in these space quanta would be $\sim 10^{80}$ GeV.

The energy associated with these stretched space quanta in three-dimensional space could be responsible for the current composition of energy and structure of the Universe. According to the SC hypothesis, this total energy is partly converted into ordinary energy, as well as into hot and cold dark matter and radiation, and the remaining into vacuum or dark energy.

To sum up, we have given somewhat plausible, though far-fetched, explanations for several unanswered questions with the help of the SC hypothesis. The SC hypothesis also seems to have some elements common to the well-accepted models and theories, such as the Big Bang model and inflation theory. Instead of seeking cancellation of most of the huge positive energy density with negative energy density, and using supersymmetry and superstring theory, the present hypothesis provides an alternate scenario. Nevertheless, it is based on sheer conjecture at this stage. We need to prove or disprove these conjectures. It does provide a new perspective that might be helpful. We shall leave it at that, lest we go further and step further into the realm of fiction.

4.6 The Future of Science

Regarding the future of science, we can think of an amusing dialogue between a 'Master' and his 'Disciple' that might proceed as follows:

Master: Why do you look so frustrated?

Disciple: The Universe still looks mysterious. Where am I?

Master: Someday, we might have the answer.

Disciple: When will it be?

Master: Don't know.

Disciple: What kind of answer do you expect?

Master: Don't know.

Disciple: Who knows?

Master: God.

Disciple: What is God?

Master: Don't know. Only God knows.!

We'll leave the issues concerning God, and move on to predict the future of science. Could anyone have predicted, a few hundred years ago, the present course of science? Predictions about the future are always difficult to make, and they are usually off the mark. As an example, we mention the very demise of science predicted by science futurists not long ago. They were predicting the demise of science after Newton, and then after the atomic model. No one could have foreseen Einstein's theories, quantum revolution, and their impact on science and technology.

No one can predict today what would become of the present scientific theories, like string theory. We cannot use science to predict the long-term future of science. Nevertheless, it has not prevented several scientists and the futurists from making future predictions in the past. It is always tempting to indulge in this exercise.

One might think that the scientific basis for predicting the future of science could be found in modern prediction theory. This theory was developed in electrical science, along with estimation and filtering techniques. Modern prediction theory for electrical signal processing is based on processing the currently available data in an optimum manner, according to an optimality creation. It then comes up with a prediction about the future signal. Unfortunately, such a theory can make only very short-term predictions with some reliability, but no good long-term predictions based on currently available data. This is because it is linear and does not take into account sudden significant abrupt future changes in data.

It is difficult to predict when exactly these events would occur. However, optimists see a bright future for science, and expect some remarkable discoveries very soon. Realistically, the only thing we can safely predict is that the future of science is unpredictable. However, the future of science would be largely dictated by the efforts to answer the remaining questions in science. Asking the right unanswered questions and seeking answers will set the future course of science.

Most of the research in science is also done based on linear thinking. It only keeps incrementally extending the well-known concepts. There have been very few lateral thinkers like Einstein, who revolutionized the field of science by injecting radically new ideas. In other words, it is difficult to predict a sudden outburst of unforeseen discoveries, which could change the entire period and direction of scientific research. As we discuss the future of science, we would also indulge in some lateral or speculative thinking.

Let us review the list of unanswered questions in science. The questions about Newton's classical mechanics and law of gravitation mainly concern the true nature of inertial and gravitational mass, causality, and the mechanisms of transferring energy from one form to another. Questions regarding the electromagnetic field mainly concern the true nature of the photon, electric charge, quanta of energy, and the mechanism of electromagnetic wave propagation in 'free' space.

Unanswered questions in modern physics concern the origin of various particles, including field carrier particles, wave-particle duality, and the uncertainty principle, because of introducing probability into science through quantum mechanics. In relativity, questions concern the constancy of the speed of light, the equivalence of inertial and gravitational mass, and the incompatibility of quantum mechanics and gravity. In cosmology, questions concern the possibility of additional particles, dark matter, dark energy, the beginning and the end of the Universe, and the nature of space, time, reality and black holes.

Sometime during the first quarter of this century, we are most likely to achieve the following:

1) Gain a better understanding of the true nature of mass.
2) Develop a better understanding of space, time, and their origin.
3) Settle questions regarding the Standard Model and the existence of Higgs particles through experiments at CERN by 2020.
4) Develop a better understanding of the supersymmetry principle.
5) Understand the hierarchy problem, the widespread gap in the energy scale between discovered particle masses up to 200 GeV, strong and electroweak unification energy scale of 10^{16} GeV and all the way up to the Planck scale 10^{19} GeV.
6) Unlock the mystery of dark matter and dark energy.
7) Develop string theory further, and extend applications of non-perturbation techniques to string theory.
8) Formulate a preliminary theory of quantum gravity.
9) Understand more clearly the true nature of quantum space.
10) Understand the role of black holes in the Universe.
11) Develop and apply chaos and fractal theories further to understand their true impact on the behavior of natural systems.
12) Put the Big Bang model and inflation theory on more solid ground.
13) Understand better the role of information science in the Universe.
14) On the genetic front, understand the function of most of the genes in human DNA, and the relationship between mutations and diseases.
15) Develop a better understanding of nanoscience and the properties of materials on nanoscale.

During the second quarter of the century, we might achieve the following:

16) Complete understanding of physics and mathematical models of phenomenon at sub-Planck scale and quantum space.

17) Complete the understanding of the true nature of time and space.
18) Develop a physical understanding of the 'quanta' of energy.
19) Resolve the conflict between relativity and quantum mechanics.
20) Develop a more general Standard Model.
21) Understand the physical mechanisms that create various fields, and discover the unified theory for all force fields.
22) Understand what happened before and immediately after the Big Bang, and predict the ultimate fate of the Universe.
23) Understand the exact role of black holes in the Universe.
24) Understand the origin and significance of the cosmological six numbers, and their values.
25) Develop a complete understanding of nanoscience and its connection to the properties of materials.

The crystal ball gets hazy beyond this point. Scientific developments are difficult to predict during the second half of the century. However, we will go out further on a limb and predict that, by the end of this century, we will have the complete answer to the question of, 'Where am I?' In other words, we would understand the true nature of this Universe - namely, how it came about, how it evolved in space and time, and what would happen to it. This would certainly not mean the end of science. By then, we would have created more questions that are even difficult to formulate at this point.

We would still have the problem of complexity, which builds complex structures from simple constituents. Ironically, most of the complex structures are similar in a particular class and yet different from each other in some ways. We observe this fact in genetic-based structures, such as trees, different animals and humans, and inanimate objects, like rocks and coastal topographies. Science has already discovered fractal geometry to describe such structures in terms of fractal dimensions. Scientists are looking at fractal and chaos theories, which can generate a lot of diversity and complexity.

Science needs to address the problem of complexity to ward off the challenge by religion that only God can create the complexity observed in the Universe and in various forms of life. There would be many unexpected turns and developments in science that would need further scientific investigations. We would still be developing better and more efficient energy conversion and propulsion techniques, new material science and materials, and unlocking the new mysteries of nature.

Money allocated for science and technology research by major countries in 2004 was only a few percent of their GDP. However, it is increasing and the research activity is intensifying. The European Parliament has approved, in 2006, a 54bn Euro (£36bn) plan to boost science research in Europe. The program is due to run from 2007 to 2013. It is interesting to note in this program that Information and Communication Technology (ICT) is the biggest winner (9.1 billion Euros), followed by health (6 billion Euros), transport (4 billion Euros) and nanotechnology (3 billion Euros). We can say one thing with confidence - that the future of science is bright as scientists march towards new discoveries, which might finally answer the question, *'Where am I?'*

4.7 The Big Picture

What does this all mean? We have unfolded a fascinating picture regarding the origin and evolution of the Universe. Upon the completion of our journey, readers would have mixed feelings. Some readers might be dazzled and impressed with what science has achieved. Others might be disappointed, since many fundamental questions remain unanswered. Regarding the origin and evolution of the Universe, the best theory we have is the Big Bang model and the inflation theory. Scientific evidence for the origin and evolution of life seems to suggest random experimentation, evolution, and survivability in a changing environment. However, creationists, and most religions, disagree and believe in the Creator.

As regards the Universe, current theories guide our understanding of the first fraction of a second, when matter and radiation exist in an extremely hot and dense form. The four fundamental force fields of nature remain unified at that time. One might understand the evolution of the Universe after that, but not the physics. We do not know the laws at work before and at the time of the Big Bang. We cannot recreate such conditions in a laboratory and must depend on theory and computer simulations. We presented some radically different approaches that stir up these issues and attempt to provide plausible answers. The sub-Planck scale physics, space and time quanta, and the SC hypotheses warrant further investigation.

We need to go way beyond the current frontiers of science and develop an exact understanding of space, time, energy, force fields, matter, information, and their interaction. We also need to come to certain definite conclusions regarding dark energy, its origin, magnitude, relation to other fields and its role in the Universe. We need to break away from conventional thinking. We need to understand the exact mechanisms through which different force fields interact.

To conclude, we must resolve the central problems in fundamental physics that still remain unresolved, namely, the incompatibility of quantum mechanics and Einstein's theory of gravitation. At present, we need two different hats to visit the Universe at the microscopic and macroscopic levels. Attempts to quantize the gravitational field in the past two decades have failed. There are still some doubts about particle physics and the Standard Model. Scientists are still chasing the goal of unifying all the forces of nature. String theorists believe that string theory can provide the answers, but others are not so sure or enthusiastic.

Chapter 5
Search for Truth about the Universe

5.1 Introduction

We are now ready to address the following question: What is the present scientific truth about our Universe, its origin and evolution, and its ingredients - namely, matter, energy, space, time, and information? However, let us first explain the meaning of the word 'truth'.

5.2 Meaning of Truth

What is the ultimate truth about the Universe, and can we ever discover it? How do we go about searching for the ultimate truth? Einstein once said that no problem could ever be solved at the same level of consciousness that created it. In search of truth, therefore, we must leave our comfort zone and search for the truth from different perspectives. We should go in search of the truth, and see how far our intellectual and experimental capabilities would take us.

As we search for the ultimate truth, we might find that we cannot discover it, because of our intellectual limitations or due to the constraints of time, space, and causation imposed on our mind. Nevertheless, we might at least get a glimpse of it, and see some light at the end of the tunnel.

To find the truth, we must first understand the exact meaning of 'truth'. Let us first examine various elements involved in discovering the truth itself. Our mind plays a critical role in deciding the truth. Since our mind decides what we accept as the truth, it limits our ability to understand the meaning of truth. According to Patanjali, there are several states to our mind:

1) Proven thinking, assumed to have been reliably proved, and thus constitutes right knowledge.
2) Unsound thinking or wrong knowledge, false assumptions, presumptions, beliefs, deductions, and inference.

3) Fancy, hallucination, or imagination totally unrelated to any proven or assumed theories, including the delusion.
4) A state of dullness or sleep, succumbing to the movement of thought, feeling it is impossible to go beyond it.

Some people even claim there is no universal, definitive, ultimate truth that stands on its own. Despite best efforts, one can only find one's own version of truth, depending on one's perception of reality.

According to an ancient Indian philosopher, Patanjali, in his book, *Patanjali's Vision of Oneness:* an Interpretive Translation by Swami Venkatesananda, proven theories derive their proof from one or the other of the following sources:

1) Direct perception, sense-experience, or intuition
2) Deduction or extension of direct perception, and sense experience or beliefs
3) In the absence of direct proof or experience, indirect proof deduction from the right or wrong application of principles of logic, chosen by one, which could often lead to vague generalizations or presumptions that since the theory comes from a usually reliable source, it must be correct.
4) Scriptural or other trustworthy testimony or authority, where one accepts as proofs the statements of those whom one has accepted as the authority. Such acceptance is faith-based, blind, and fanatic.
5) Unsound thinking or wrong knowledge based on error, and on mistaken identity. In this case, the cognition is unreal and faulty and hence the knowledge is faulty, too. There is no agreement between the expression and the experience, between the substance and the description.

We define scientific 'truth' as one that is objective, which a scientist can verify directly through experiments, or conclude indirectly through logical thinking.

5.3 Universal Principle of Change

A truth that stands alone is the Universal Principle of Change. Surprisingly enough, the Universal Principle of Change always holds in this Universe. It governs the evolution of the Universe, life, and even our behavior. In this dynamic Universe, all inanimate entities, involved in various phenomena, follow the Universal Principle of Change (UPC). Why is change eternal, perpetual, and why is the Universe dynamic? One can trace back the ultimate cause for this to the Big Bang, which triggered the motion and expansion of space, and led to the splitting of the energy field into four force fields. Since then, the process of change continues, as various entities keep on seeking the elusive equilibrium.

Different fields interact, maintaining the energy balance at every instant and transforming one form of energy into another. When different fields interact, there is an action and a reaction. The so-called 'cause' or the associated field interacts with another field in the region. The acting field affects the other field(s) in the region, and the other field(s) affects it. What we observe is so-called effect, because of this interaction that maintains the energy balance through transformation and exchange, following the path of least resistance (minimum work principle).

Suppose a ball is resting on a rough surface, and we apply a certain force to it. What actually happens? The matter has no intelligence, and it must obey and follow the physical laws of nature. The applied force generates distortion in the field of atoms of the ball in contact with the surface, resulting in the modification of the electromagnetic field. The subsequent action follows the principle of least action. Part of the applied energy transforms first into heat, generated at contact (friction). Depending on the direction and the magnitude of the applied force, it might transform all the energy to heat, and not move at all. Alternatively, the force field displaces atoms, and thus the entire ball, in space, according to the laws of motion. However, the Universal Principle of Change embedding the principle of least action is always in play, and the total energy during the transformation is always preserved.

The material entities that follow the Universal Principle of Change include fundamental particles, material objects, planets, stars, and galaxies. Robust structures or stable outcomes keep on surviving, following the path of least resistance (minimum work principle), and others simply perish. As quantum mechanics tells us, nature introduces an element of uncertainty and randomness, and an object can exist in many states simultaneously. However, the moment we observe, these possible states collapse into one outcome as the time comes into play. We observe only one choice, and think that this is the reality and that there was no other choice.

All living entities also follow the Universal Principle of Change, as they adapt to sudden energy or environmental changes. However, they do have some choice and can usually choose an optimality criterion rather than follow the path of least resistance (minimum work principle), depending on their level of intelligence. They project their individual personalities, and have some conscience or awareness of the Universe and so-called intelligence because of the neural connections in their brains.

Their decisions and subsequent actions to adapt to environmental changes are dictated by the patterns they store in their brains. They observe through the senses, process the information with the help of the brain (intelligent computer), and evaluate it against stored patterns or values. They then follow the decision they consider best or optimum under the circumstances, i.e., subject to the constraints of their physical and mental capabilities, and the goals they set for themselves according to their acquired values.

While searching for scientific truth, let us examine, one-by-one, the various ingredients that constitute our Universe.

A. Truth about Ordinary and Dark Matter

Science has already discovered the basic truth about ordinary matter and radiation, which constitutes about 4% of our Universe. All ordinary matter is composed of fundamental particles, called electrons and quarks. The elementary particles electrons revolve around a nucleus of an atom. The protons and neutrons inside the nucleus of an atom are composed of quarks. It is the number and arrangement of such particles that differentiates atoms of one element from another.

These atoms also combine to form molecules of compounds, which form a large part of ordinary matter. However, we do not understand the origin of the fundamental particles. Nor do we understand why this particular generation of particles was chosen for our Universe.

Matter can also exist in other forms. For example, a cloud of sodium atoms trapped in a vacuum and cooled to just above absolute zero (-

273°C), coalesce to form a Bose-Einstein Condensate (BEC). One can produce BECs usually by super-cooling atoms so that they merge and begin to behave like one giant atom. The BEC, an exotic quantum entity first predicted by Albert Einstein, was created in the lab in 1995.

One can bring light to a standstill in this condensate. The Harvard team, in the late 1990s, slowed light from its constant 299,792km/s (186,282mps) to a mere 61km/h (38mph). It applied the brakes by shining light into a cloud of sodium atoms trapped in a vacuum and cooled to just above absolute zero (-273°C). A second laser tuned the tiny atomic cloud to slow the pulse of light. In 2001, working with a team from the Harvard Smithsonian Center for Astrophysics, the same group brought light to a halt, by slowly turning off the second control laser. Switching the laser back on set the light free.

Science has yet to discover the ultimate truth about dark matter, which is about 6 times the size of ordinary matter and constitutes about 23% of our Universe. It would appear that such matter should also be composed of particles, but we have yet to discover such particles. Many scientific theories predict such particles, and experiments in high-energy accelerators, such as at CERN, are underway to find such particles.

We cannot detect dark matter directly, but researchers note its presence indirectly. Anything that has a mass exerts the force that we call gravity. According to scientists at NASA's Goddard Space Flight Center, dark matter exerts a gravitational pull on objects in and around distant galaxies, and on the light emitted by those objects. By measuring these mysterious effects of gravity, researchers determine the amount of 'extra' gravity present, and hence the extra mass, or dark matter, which must exist.

In large clusters of galaxies, scientists say that 5 to 10 times more material exists than can be accounted for by the stars and gas that they find. To reach us, light from the galaxies has to pass through intervening dark matter. This dark material bends light in much the same way as it is bent when traveling through a lens. The deflected light distorts the shape of the background galaxies. Thus, we see them in a distorted way, as if through lots of little lenses - and each of those lenses is a bit of dark matter.

Richard Massey and his colleagues used this weak gravitational lensing technique to detect the dark matter. They published in the journal, Nature, a 3-D map of the dark matter mass distribution. It was based on 1,000 hours of observations through the Hubble Space Telescope.

The mass distribution in the map is based on measurements of about half a million distant galaxies. The map shows that the concentrations of ordinary matter usually overlap with concentrations of dark matter - but not always. Conversely, the dark matter concentrations sometimes seemed to have no corresponding ordinary matter, or galaxies. The map seems to confirm the cold dark matter theory for cosmic evolution.

According to this theory, soon after the Big Bang, cold dark matter formed the first large structures in the Universe, which then collapsed under their own weight to form vast halos. The gravitational pull of these halos sucked in ordinary matter, providing a focus for the formation of galaxies.

Recent observations of a great big cosmic collision also provided evidence for the existence of invisible and mysterious dark matter. This collision between two huge clusters of galaxies was the most energetic cosmic event besides the Big Bang. The researchers discovered essentially the gravitational signature of dark matter. This signature was created by dark matter and ordinary matter being wrenched apart by the immense collision of two large galaxy clusters. The impact split normal matter and dark matter apart, rendering dark matter's gravitational signature observable. According to Doug Clowe, from the University of Arizona, this provides the first direct proof that dark matter must exist, and it must make up the majority of the matter in the Universe.

In 2007, the gravitational lensing also allowed researchers from Johns Hopkins University and the Space Telescope Science Institute to spot a ring of dark matter unexpectedly, while they were mapping the distribution of dark matter within the galaxy cluster Cl 0024+17. This cluster lies 5 billion light years from Earth; its ring of dark matter measures 2.6 million light years across. At first, team members thought the ring was an illusion - or artifact - in the data. However, repeated experiments confirmed its existence.

B. Truth about Energy

Science has discovered four force fields: gravity, electromagnetic, weak, and strong fields. These fields store different forms of 'ordinary' energy. Scientists have developed theories to unify all these fields except gravity. They have also discovered fundamental carrier particles for these fields, except gravity. String theory predicts the existence of graviton – the carrier particle for gravity - and experiments are underway to find graviton.

Science has also discovered that energy associated with these fields has its own currency, and its lowest denomination is a quantum of energy. Science cannot explain the origin of all this energy, though it can explain the interaction between the different forms of energy. Scientists are also completely in the dark about dark energy, which constitutes about 73% of our Universe. They have no idea about its origin, its carrier particles, or any of its properties.

Scientists claim that this dark energy is responsible for the expansion and acceleration of the space in our Universe. This mysterious force, speeding up the expansion of the Universe, has been a part of space for at least nine billion years. That is the conclusion of astronomers who presented results in 2006 from a three-year study using the Hubble Space Telescope.

Using NASA's Hubble Space Telescope, scientists measured the expansion of the Universe 9 billion years ago, based on 23 of the most distant supernovae ever detected. There is lot of speculation, and scientists have advanced several hypotheses about dark energy. However, science has to go a long way before it can discover the true nature of dark energy.

Based on a little evidence, one might speculate as follows. Science would find carrier particles for each form of energy, just as it has discovered fundamental particles for matter. Just like matter, it is likely that science would also, some day, discover a unified theory, which would unify and tie together the different forms of energy and the associated force fields.

It seems that various forms of energy are a manifestation of the one field, and different force fields and the carrier particles are a manifestation of this one field. Under certain conditions, different carrier particles and the corresponding forms of energy come into existence. However, at this stage, this is speculation, and we are far from discovering the ultimate truth about energy.

C. Truth about Space & Time

Thanks to Einstein, we now know that space and time are intertwined and inseparable. Space and time come into play when we observe an event. We label an event by a series of numbers. Three numbers tell us where it happened in space. One tells us when it happened in time. We measure these four numbers with measuring devices, such as measuring rods and clocks.

According to Newton, the properties of accurate measuring rods and clocks can be made completely independent of the system. However, Einstein showed that this is not the case. Two observers in motion, relative to each other, do not make identical measurements of space and time, but these measurements are related.

Einstein's special relativity theory, using the invariance of the speed of light in space, concludes that a space dimension would shrink for an observer and time dilates for observers moving relative to each other. This is due to the simple fact that the speed of light (=distance/time) remains constant in free space, which implies that the measure of distance and time for two observers moving relative to each other has to change. Hence, the measure of time and space is not absolute.

Simultaneity, time, and space are all relative, since it depends on the observer's frame of reference and his relative speed with respect to the observed reference frame. If an observer could move at the speed of light, relative to a frame of reference, the observed length would shrink to zero and time would stop for the observer.

Is space empty and inert? Is space a container or a dynamic entity? Scientists first suggested that space is filled with ether. Then it was declared empty after experiments failed to detect any variation in the speed of light. Einstein's general relativity theory suggests the curvature of space as a measure of gravity. Gravity alters space and the resulting deformation of space is responsible for the gravitational force.

Einstein's relativity theories have thus demolished the conventional notion of absolute space and time. Now, scientists also suggest the quantum nature of space, where virtual particles keep on appearing and annihilating each other. It is suggested that space is not a continuum, and it is made of Planck space quanta. Each quantum of space has the following parameters associated with it: Planck scale length l_p =1.616×10^{-35}m, time t_p =5.3906×10^{-44} sec, with energy E_p = $1.2212 \times 10^{+19}$ GeV, and energy density ρ_p = $4.64 \times 10^{+113}$ Joules/m^3.

Is time like a river, flowing only in one direction? Let us see what science has to say about this question. We observe events occurring in space, or note a change in some quantity, because of the interaction of different force fields and the resulting transformation and exchange of energy. Time is a convenient parameter to record an effect due to a cause. An observer can record the changes in the value of a particular quantity in an ordered sequence.

Time thus comes into play, usually when we observe an event. Time, for an observer, moves only in one direction, unlike space. One can only observe the present, but one cannot go back in time to the past. With the passage of time, the future is revealed to us, but we cannot foresee the future at any particular moment. It is interesting to note that, if a person loses memory and his mind cannot register change, the concept of past, present, future, and of time itself, disappears.

Some scientists suggest that if space is discrete and we can only define the smallest length as the Planck length, then time must also be discrete. We use Planck time (t_p =5.3906×10^{-44} sec) as the quanta of time. Time does not seem to exist by itself in quantum space. The notion of time arises only when space contains energy or when a particle (a singular energy configuration) moves with a certain delay among space quanta.

In other words, time is associated with the particle position or energy movement or change. Since the speed of light (c = l_p/t_p) is finite, energy or matter at one point in space cannot appear at another point in space at the same instant, because it cannot transfer instantly. Thus, space imposes a certain delay on energy as it travels through it. The particle, or the singular energy configuration, reveals itself as it moves or transfers from one discrete space element or space quantum to the next. We identify this delay for the successive transitions in quantum space from one to other space quanta, as the quanta of time.

Have space and time always existed, or did they originate when the Big Bang occurred? Scientists, advancing the Big Bang theory for the origin of the Universe, suggest that space and time were created when the Big Bang occurred. However, they have been unable to tell us what happened during the initial moments of creation, before 10^{-43} sec. They have not been able to develop any credible scientific theories or physics below the Planck time scale. Furthermore, science cannot confirm if there was anything before the Big Bang, and what caused it. Some scientists have recently advanced theories that project a scenario about the pre-Big Bang era.

Despite these unusual results by Einstein and other scientists, scientists still do not understand the true nature of space or time. It seems that if there were no matter and space, time would not exist. Furthermore, in transcendental meditation, when the mind is bereft of any thought, it loses sense of space, time, and causality. It steps into a different domain, called the 'domain of God' by the spiritualist. The Universe does exist and keeps on changing, whether someone is there to observe it or not. In my opinion, the ultimate truth concerning space or time is yet to be discovered, and scientists need to develop different concepts.

D. Truth about Information

Scientists have discovered that information is also an essential ingredient in the Universe. The Universe and life has complex information codes, which describe everything that happens. Regarding our Universe, Wheeler said in a lecture,

"Every particle, every field of force, even the space-time continuum itself derives its function, its meaning, its very existence entirely from binary choices, bits. What we call reality arises in the last analysis from the posing of yes/no questions."

Some physicists are suggesting that when we hit a wall, we knock into information. These scientists claim that particles, like electrons, ions, and atoms, have certain properties. If we can transmit this information and reproduce the properties of the quantum particles making up an object in another particle group, we could precisely duplicate the object. In other words, we need to transmit only the information about the particles' properties and not the particles themselves.

Some biologists have proposed that life itself might be information. Olson, in a recent lecture, said that the most wonderful achievement of the century has been the marriage of biology and computers. This characterizes the entire human genome in terms of the bits of information representing A, C, T, and G sequences. In life sciences, we can put the entire 3 billion digits of our DNA on about four CDs at the current rates of compression.

This 3-gigabyte genome sequence represents the prime coding information of a human body, i.e., our life as numbers. It implies that biology can be conceived by science as an information process, and it shows that life itself is information.

Regarding the ultimate truth about information, scientists have recently decoded the information about life contained in a DNA molecule. Scientists have suggested that even hard matter is information and the entire Universe acts according to a certain code. However, science has yet to decode the information code about the Universe. The physical laws discovered by science, thus far, are just the tips of this iceberg.

5.4 Truth about the Universe

Until recently, scientists thought that the Universe was essentially static and unchanging in time. We liked to believe that the Universe and the human race have existed forever. Scientists also liked the idea, because a beginning might imply the existence of a Supernatural Being, who created the Universe. This argument, about whether or not the Universe had a beginning, persisted until the 20th century.

The observational evidence has now confirmed that the Universe is not static. It is certain that something happened around 13.7 billion years ago, which brought into existence the Universe that we know now. The time scale of the Universe and its evolution period (13.7 billion years) has been much longer than life on Earth (3.5 billion years).

The Universe is expanding and galaxies are moving steadily apart from each other. The origin and evolution of our Universe, according to many scientists, seem to follow the path of apparent random perturbations, similar to the evolution of life. A scientist might develop the following scenario for the evolution of the Universe.

- Origin & Evolution of the Universe

Based on previous discussions, the following scenario can be suggested for the origin and evolution of the Universe. In the beginning, there is a space quantum (where did it come from?). Violent fluctuations in quantum space generate more space quanta, energy, and matter from the 'vacuum'. Due to the quantum mechanical uncertainty, the vacuum becomes unstable, and tiny bubbles begin to form in the vacuum, sort of the way bubbles form when water begins to boil. Each bubble expands rapidly and represents a Universe.

The string theorists call the set of bubbles, a 'Multiverse'. Our Universe is one of these bubbles. As the bubbles expand after the Big Bang, different bubbles acquire different values for the six cosmological constants in a random fashion, according to the Universal Principle of Change.

These numbers decide the fate of different universes in the Multiverse. Although a large number of bubbles form, most of them simply perish since they cannot achieve a dynamic equilibrium in the changing environment. Only a few bubbles survive and continue to evolve. No net energy is required to create the Multiverse from 'vacuum', since instability creates equal amounts of positive and the negative energy. The energy balance in the beginning, as well as through evolution, is maintained continuously, as the evolution follows the Universal Principle of Change.

Our Universe - one such bubble - went through apparent randomly varying stages and emerged the way it is today. The six cosmological numbers guided the evolution process for our Universe. These numbers had to be within an extremely narrow range. Any large deviation from these values would not lead to the Universe that exists today. Protons would decay, nuclei would become unstable, DNA could not form, carbon-based life could not happen, and we simply would not be here. The anthropic principle thus states that we observe our Universe the way it is because we exist. If conditions were any different, no one would be here to ponder them.

One might ask a related question. Is there only one set of initial conditions that can produce a Universe that we observe? Or, there exist a number of different sets of initial configurations exist? Scientists have not yet answered this question. According to Hawking, cosmologists are also investigating whether initial configurations favored by the no boundary in space or time, coupled with anthropic arguments, could lead to the evolution of our Universe.

The most likely scenario would seem to be similar to the evolution of life on Earth. The emergence of the existing Universe is the result of a long, random, dynamic chain of events and perturbations, driven by the Universal Principle of Change.

Incidentally, a typical example of random perturbation due to a changing environment would be someone unexpectedly opening a door of a room to the outside, changing the temperature inside the room. Other so-called random perturbations may be due to the sensitivity of a nonlinear system or phenomena to initial conditions or chaotic behavior, discussed in the recently developed theory of chaos.

There also remains the possibility of the evolution and existence of other universes with a different set of six numbers. We have said that our Universe, and the life we know, could not exist if six numbers differed significantly from present values. However, there could be different universes with different sets of physical laws and different sets of constants supporting multiple forms of life, different from our form of life.

Some might have little or no gravity, others stronger gravity. Some might be stable and long-lived; others might be unstable and short-lived. It is highly unlikely that we would meet such universes. New Universe could also be triggered within the interior of a black hole when black holes collapse, creating space and time that is disjoint and never overlapping our space and time.

No matter how the Universe originated, the following important question remains unanswered. It is a common belief that our Universe had a beginning. However, an event is caused by an earlier event, and the chain continues. We can always ask, 'What caused the first event in the chain?' On the other hand, could this chain be closed, and thus, have no beginning or an end? Science has so far avoided answering such questions. Hawking says,

"We must try to understand the beginning of the Universe on the basis of science. It may be a task beyond our powers, but we should at least make an attempt."

In an interesting public lecture, "The Beginning of Time", Hawking argues that the Universe, and time itself, had a beginning in the Big Bang, about 15 billion years ago. The Universe that we know has not existed forever. He also introduces the novel idea of an imaginary time in addition to the real time. The notion helps us to avoid specifying boundary conditions if the histories of the Universe in imaginary time are closed surfaces, like the surface of the Earth.

- Outside Intelligence & the Universe

How did the laws of nature originate for the origin and evolution of the Universe? Were there any outside intelligence, preconceived plan or intelligent design, and a one-time or continuous divine intervention in the origin and evolution of the Universe? The preceding discussion seems to suggest that there was no inherent built-in design to create the Universe as it exists.

It would appear that a bubble did not follow a deterministic path to emerge as the present Universe, because of the random perturbations in a complex changing environment driving the evolution. The evolution path appears to have resulted from a series of coincidences and not from carefully chosen events in advance. According to this line of reasoning, there was no intelligent, built-in, or preconceived design for the Universe.

However, different people interpret the available evidence differently. Agnostics and many scientists choose to neither believe nor disbelieve in God or in Intelligent Design. They would like to wait for more information, since all the related questions have yet to be answered. For example, it could be argued that what we call random is not random, but a highly complex deterministic sequence of actions according to certain laws that we do not yet understand.

The atheists, who do not believe in God or in Providential Intervention, believe the evolution to be a sheer coincidence, and they point to the role of chance in evolution. The believers of the anthropic principle say that, since we exist in such a Universe, we are able to ask the question of,

'Why it is this way?' If our Universe had evolved to a different state with life forms different from ours– or if other universes exist – then different types of living beings would be asking the same question.

In short, the available scientific evidence, though not complete, does seem to suggest that our Universe evolved dynamically through an apparently random selection process, following the Universal Principle of Change and the associated laws of nature (origin unknown!). It seems to suggest that our Universe, and the life in it, evolved because of random interactions. It proceeded along the paths that could survive in the prevailing environment, amongst various outcomes, because of random experimentation.

Based on our current understanding of the available evidence, one might be tempted to make the following statement: Suppose our Universe and life were to start again, then it appears unlikely that either of them would evolve to the current form. Some of the random factors, such as the environment or the random perturbations and selections involved, might change during the evolution process. For the present Universe to evolve, too many events had to occur in a particular sequence and at certain times.

The theists, who believe in God or in continuous Providential Intervention, interpret the available evidence differently. They believe in a pre-destined Universe, created by God and guided through His physical laws. They consider randomness as a substitute for our ignorance and lack of understanding of the underlying phenomenon due to its vast complexity.

The deists, who believe in God's one-time Providential Intervention, point to the one-time intervention at the time of Big Bang, and the physical laws, as set initially by God. Their argument in favor of outside intelligence is the existence of such intelligent laws of nature responsible for the origin and evolution of the Universe. The proponents of Intelligent Design want to know: how did such laws come into existence?

Religious thinkers try to interpret scientific data to support the need for the Creator and the existence of God. Thus, the Universe needs a Creator. According to religious belief, the same laws that the Universe has to follow, of course, do not bind the Creator. The Creator is described as the creative entity transcending space, time, and matter and does not require creation. Believers do recognize that they cannot prove the existence of God. Nonbelievers argue that an uncaused, un-designed emergence of the Universe from 'nothing' does not violate any principle of physics. The total energy of the Universe is always zero, so the Universe did not require a miracle of created energy from 'nothing' for its origin.

How did the 'intelligent' laws come about for the origin and evolution of the Universe? Science cannot yet answer many such questions. Of course, science cannot and should not substitute mere opinions just because it has not yet found answers to all the questions. The process of scientific inquiry should continue to search for the remaining answers. We shall discuss the outside intelligence, Intelligent Design theory and different intelligent life forms from scientific and religious perspective in the third phase of our journey.

To sum up, science has yet to discover the ultimate truth. Science has yet to discover a great deal before we have the complete picture. However, we need scientific theories and not speculation. Hawking's recent book, The Universe in a Nutshell, attempts to look inside the Nutshell, but this shell is hard to crack yet.

In due course, hopefully, scientists might put all the pieces of the puzzle together and come up with the final answer. This final answer would crack the Nutshell, and reveal the ultimate truth. Only when science answers all the remaining unanswered questions, can it can celebrate its final victory! We shall have more to say on this subject in the last chapter of the book.

Chapter 6
Final Destination - The Ultimate Truth

6.1 Introduction

The journey of science has indeed been quite remarkable. Scientists have come up with physical laws that explain different phenomena occurring in nature. Science can explain how our Universe evolved, and how various elements formed, starting at 10^{-43} seconds from the so-called Big Bang.

As we come to the end of our journey, we realize, however, that science does not have all the answers about the Universe. In science, for example, we do not know the origin of the energy for the Big Bang. We feel uncomfortable with randomness in quantum mechanics. The incompatibility of quantum mechanics with relativity theories is also troublesome. We also do not have a satisfactory explanation for most of the energy in this Universe. We do not understand dark energy and dark matter, which are required to explain the present structure of the Universe.

We must remember that the Universe does its own thing, whether we understand it or not. Science does not invent new things; it merely discovers the existing facts. Newton and Einstein merely uncovered laws, which were already in existence. Another important point to remember is that scientists coin certain terms - such as energy, force field, matter, space, time, information, change - and use them to describe certain observations. Nature does not use such terms to define its processes.

6.2 Human Limitations

We must first understand our limitations, and accept the fact that we might have a flawed interpretation of the reality of the Universe. When we talk about the material world, or the Universe, we are looking outwards from the inside of our brain. We thus want to look at the form and the structure of the entire Universe. It is as if someone, locked inside a bottle, is trying to read the label on the outside of the bottle. Alternatively, someone locked inside a room is trying to know what the room looks like from the outside. Unless one can step or project outside, one cannot get the complete picture.

We are part of the whole, and the part wants to understand the whole. It is like a tiny part of our body trying to understand the working of the entire body. However, some scientists believe that our Universe is like a hologram. In a hologram, the entire form is enfolded within each part, and each part contains enough information to reconstitute the whole. If this premise is true, we do have a chance to understand the real Universe

The most important difference between a human being and rest of the animals is the human's brain, which developed during the process of genetic evolution. It contains about one trillion cells with 100 trillion connections between those cells. Although we use less than ten percent of the neuron cells, they can make very complex connections resulting in complicated patterns.

Our brain is a fast, large and highly intelligent computer, but it has limitations. It can process and interpret data only to a limited extent, observed by our limited sensors. Our intelligence is finite, and it would appear that the finite could not understand the infinite. Our brain is also a prisoner of time, space, and causality.

Thus, our perception of reality and the Universe is not absolute. Our attempt to perceive the true nature of this vast Universe, through the limited senses and interpretation of the data, seems like several blind persons trying to visualize an elephant holding different parts. Even the terms, such as time, space, etc. that we define to express our understanding of the Universe need better definitions and generalization. Unless we understand how our brain understands, we cannot truly understand this Universe.

Regarding our brain, despite the marvelous success of neuroscience in the past century and developments in artificial intelligence, we are far from completely understanding the cognitive processes. The basic difficulty lies in understanding various connections of neurons, the formation and identification of patterns, and the brain signal processing mechanisms. These are collectively responsible for cognitive activity and decision-making. Our brain processes data and comes up with decisions, based on our values and compatible optimality criteria.

One cannot model the brain as a simple cause-effect model because of the large number of connections. On one hand, our brain makes decisions based on fuzzy logic, without solving all the dynamic equations, when we drive a car in traffic. On the other hand, it is capable of understanding the most exact logic.

Scientists are trying to gain a better understanding of the brain. They are also attempting to build intelligent computers with the capability to make complex connections like the ones that neurons make in our brain, and to process the data 'intelligently'. If that happens, we might be able to amplify our intelligence, just as the other machines amplified our mechanical power, and get a better picture of the Universe.

6.3 What We Know About the Universe

Let us collect our thoughts together in one place about the Universe.

- Interesting Features of the Universe
1) Everything in the physical Universe is interdependent, interrelated, and interconnected, forming a vast and complex Universe.
2) Nature keeps on doing what it does, whether or not someone is there to observe it or understand it. Causality, cycles, and change are the main characteristics of all phenomena in this Universe.
3) We usually create partitions or compartments to reduce the complexity. For example, science considers different fields individually, and thus restricts itself by not considering all the interactions at once. Newton's law explains, in a simplistic manner, the motion of an object in terms of mass, force, and acceleration. It does not explain the complex field interactions that occur between different fields.
4) An observer opens a window to the Universe through his sensors and tools. Based on observed data, processing

capabilities, and the terms he defines, he expresses his understanding of nature. Scientific discoveries are also limited by the range of tools used for measurement (e.g. one cannot measure length or time anywhere close to Planck scale), and by our mental ability to analyze the problems. In fact, we are part of the problem being an integral part of the whole system. Unless we can step outside this box, we cannot get the complete picture

5) We assign various properties to space and matter without a clear understanding of their real nature and the origin of such properties. We run into difficulty when we examine extremely small distances and time (below Planck scale) as the notion of separate space and time does not hold.

6) In terms of our definitions, dynamic field(s) with associated energy (agent for change) propagates in waves continually, and permeate the entire Universe. The sine wave is the most fundamental wave, since one can express any field waves in terms of sine waves.

7) Matter is a manifestation of energy fields - a composite of singular energy configurations (bundles of energy with different attributes) called fundamental particles. Different fields are also composite of singular energy configurations – called carrier particles.

8) The Universal Principle of Change governs all dynamic phenomena resulting from the interaction of different fields and/or matter, and continually seeking equilibrium following the principle of least action. During change, it conserves total energy and total momentum.

9) All the discovered physical laws of nature explain the change in this physical Universe as cause-effect relationships, using terms such as time, space, matter, atoms, gravity, etc. The Universal Principle of Change seems to form the basis of the laws.

10) The complexity of our Universe and nature is awe-inspiring and overpowering. Our intelligence is no match for it. Our finite intelligence is simply inadequate to answer the most

fundamental questions about the Universe and the complex interaction of force fields, and to explain completely even a simple phenomenon.

- Top Ten Mysteries of Science –Comments

Let us now return to the top ten mysteries of science, and discuss the most likely resolution of these mysteries. It is worth mentioning at the outset, however, that we do not have any definite answers yet. Nor do we have any satisfactory theories to explain various aspects of the questions. Nevertheless, we shall give some plausible explanations.

After all, every scientific discovery starts with some sort of belief in an idea (in this sense, science is a religion), which is accepted if proved correct and is independently verifiable, or else, it is discarded (that is where science digresses from religion). We list the top ten mysteries, along with personal beliefs, as follows:

1) *If there is a cause for every effect, then what caused the Big Bang? Where does this cause-effect chain end? Put succinctly, does the Universe have an origin or end in time and space?*

Stephen Hawking says that this question is meaningless; it is like asking what is north of the North Pole? However, the very premise of science is searching for the cause-effect relationships. We must search for the cause and mechanism of creation, and ask whether the Big Bang can occur again. For continuity, we might have to assume that there exists a pre-Big Bang scenario. In this case, the end of the Universe might become the beginning of the Big Bang. In other words, the final effect in the Universe becomes the cause of the Big Bang. Thus, the cause and effect chain closes on itself.

2) *The interaction of the force fields and the resulting energy transformation, following the laws of nature, is responsible for all change and for the existence of this dynamic Universe. What is the origin of these force fields and the associated energy, and which transformation takes precedence?*

Scientists say that the three force fields merge at 10^{16} GeV and split below this level of energy. Gravity merges at 10^{19} GeV. Thus, a likely scenario would be the emergence of gravity, as the energy falls below

10^{19} GeV, and the emergence of the remaining force fields as the energy falls below 10^{16} GeV. As to the origin of the single cosmic field, and why and how these fields split from this single field, the answer is most likely to come from the studies at Planck scale, in quantum space, and symmetry violations.

3) *The contents of the Universe are 4% ordinary matter including radiation, 23% dark matter, and 73% dark energy. What is the exact nature of dark matter and dark energy, and do these percentages vary over time?*

As for the last part of the question, these percentages do vary, since the matter was formed much later during the evolution of the Universe. Regarding the nature of dark matter, a plausible answer would seem to be the existence of heavy particles. However, we have not yet detected such particles, simply because they are hard to detect. Understanding the exact nature of dark energy is most likely to come from the study of the properties of space, especially at the quantum level.

4) *If every galaxy has a black hole at its center, what is the true nature and role of these black holes in the formation of the stars, galaxies, their evolution, and in the fate of the Universe?*

Science has confirmed the existence of black holes. However, we do not understand their exact role in our Universe. It would appear that the black holes have been forming and disintegrating throughout the evolution of the Universe. Understanding their exact role is most likely to come from the development of quantum gravity theory – the behavior of the gravitational field at the quantum level

5) *If just six numbers are responsible for the existence of our Universe, then how were their values set? Can another Universe exist with a different set of values?*

We know that any significant change in the values of these numbers would not support our Universe, and might lead to different universes. The particular set of stable values, corresponding to our Universe, appears to have evolved following the Big Bang. The exact understanding is likely to come from understanding the physics in the Planck era within 10^{-43} seconds

6) *How and why do the Planck constant, the speed of light in free space, and the gravitational constant have the values they have? Can we derive these values directly from theory?*

The answer to these questions is most likely to come from a better understanding of the properties of space and time. For example, Einstein's gravitational theory defines gravity in terms of space-time curvature, indicating that space is not just an entity, defined for convenience to measure distances. It is a physical and dynamic entity with certain properties, such as permeability and permittivity, which relate directly to the speed of light.

7) *The Standard Model of particle physics has enjoyed phenomenal success. However, if we have discovered all the force fields, and the Standard Model still does not include gravity, then how do we generalize it and develop a grand unified field theory?*

The Standard Model relies on Dirac's relativistic equation of quantum mechanics, which ignores gravity. Therefore, it should not surprise anyone that the Standard Model cannot account for the gravitational field. We need to generalize the Standard Model and integrate it with general relativity theory. Standard Model Extension must include gravity in the model, and explain the emergence of the great variety of particles that emerge from high-energy accelerators.

8) *If the deterministic relativity theories can explain the Universe and gravity at the macro-scale, and quantum mechanics all the phenomena at the micro-scale, then how do we integrate them?*

Despite claims to the contrary and the tremendous success of quantum mechanics, could it be that Einstein's intuition makes sense and quantum mechanics is too paradoxical - a substitute for our ignorance about the phenomena at the quantum level? Nevertheless, the success and confirmation of both theories implies that both these theories are part of a more general theory that's not yet discovered.

9) *If string theory, with its recent discoveries - dualities, application of quantum mechanics and extensions to multi-dimensional 'branes'- is really the physics of the 21st century, then how do we confirm that it describes reality?*

String theorists claim that string theory will most likely unify both relativity theories and quantum mechanics. They talk about the quantum strings and 'branes' in a multi-dimensional space. They try to explain the reason for the gravity field being so weak, and even predict the existence of 'graviton' – the carrier particle for the gravitational field. It seems reasonable to expect the Planck scale (string theory uses the scale for quantum strings) to play a critical role in understanding the properties of space. However, string theory would remain only an elegant mathematical exercise, unless scientists can find some experimental evidence to support it. The limitation of our instruments makes it a very difficult problem.

10) *If our brain is limited in its computational and analytical abilities, by its hard-wired evolutionary programming, and by the constraints of time, space, and causality, then is it possible for the human mind to discover the ultimate truth about the Universe?*

We must realize that we mortals have limitations, as we are an insignificant part of this Universe. Intellectual and other resource limitations will most likely to prevent us from discovering the ultimate truth about this complex Universe. However, it should not prevent us from searching for the answers. We have come a long way in just one hundred years. We still have a long way to go, but the journey must continue. Our intelligence may be limited, but the flight of our imagination is limitless. Admittedly, science has not yet found the answers to all these questions. Being an optimist, I believe that scientists would continue to march forward, filling in the present gaps in our understanding.

- Parting Words

When we started the journey, we wanted to answer the following question regarding the material world, *'Where am I?'* To look for the answer, we visited science. However, we did not find a complete and clear answer. Scientists are still in search of the ultimate truth about the

Universe. Despite all the problems with the world, however, we have a bright future ahead of us. Human intelligence will keep on evolving, and we shall discover and conquer many new frontiers.

As far as the Universe is concerned, we live on a very small planet in the solar system in a giant Milky Way Galaxy, which is one of the billions of galaxies in the visible Universe. The Universe is vast, always changing, and its dynamics is indeed complex. We are part of this vast Universe, locked inside this box, bound by the constraints of space, time, causality, and by the limited range of our sensors. Unless, we can somehow remove some of these constraints, we have little chance of succeeding.

Our attitude towards discovering the answers about the Universe is similar to a fish caught inside a net, thrown into the ocean by an angler. Some fish are happy inside the net, and they never try to find the net, let alone escape. Others are curious and go near the net, but do not make effort to escape. Still others realize their bondage, and try to do their best to escape without success. Finally, there are a few fish that do break the bonds, escape, and swim freely into the vast ocean.

We can categorize humanity in the same manner. I have friends who do not care to know the answer to any questions about the Universe. They just want to enjoy it. I have other friends who are curious, but are satisfied with the simple answer that God does everything. Then, I have friends, mostly scientists, who constantly question and struggle to find the answers, though they do not succeed in finding all the answers. However, I have yet to meet someone, who knows the absolute truth and the answers to all the questions about the Universe.

It seems that, as long as we are constrained by our intelligence, space, time, and causality, we might not find all the answers. Science might broaden its scope, studying thought processes and the functioning of our brain to understand its limitations. Some scientists, in search of truth, might even step into the realm of spirituality to search for the existence of the soul, believed to be beyond the mind and not bound by space and time.

I hope tthe readers enjoyed reading this book as much as I enjoyed writing it. I have included several new ideas and speculations about the future development in science to answer the unresolved problems, which were discussed in the book. Whether any of these help in resolving the current problems, only time will tell!.

For any feedback and comments, please connect with me on Facebook or my website: www.knowingtheunknownbooks.com

Glossary of Terms Used

Absolute zero

Temperature of 0 K (Kelvin) at which all molecular activity ceases and a substance contains no heat energy.

Acceleration

Change in speed or direction of an object with time, i.e., rate of change of velocity.

Albert Einstein

German-American physicist; he developed Special and General Theories of Relativity, which along with quantum mechanics is the foundation of modern physics.

Antimatter

Matter with same gravitational properties as ordinary matter, but with opposite electric and nuclear force charges.

Antiparticle

A particle of antimatter; every ordinary matter particle has a corresponding antiparticle, which annihilate each other during collision, transforming into energy.

Atom

The basic unit and fundamental building block of matter, consisting of nucleus containing protons and neutrons, surrounded by orbiting electrons.

Atomic number

The number of protons or electrons in an atom

Big Bang

A theory of evolution of the Universe; assumes origin of the expanding Universe from a singularity with an explosion billions of years ago. The Big Bang theory successfully explains the expansion of the Universe, the cosmic microwave background spectrum, and the origin of the light elements.

Bit

Basic unit of information, 1 or 0, which corresponds to Yes or No.

Black Hole

A region in space in which gravity is so strong, that nothing, not even light, can escape from it. Current indications are that there exists a black hole in the center of every galaxy.

Boson
A particle with integer spin, which is typically a carrier particle for a field.

Brane
Extended fundamental objects in string theory; 1-brane is string in one dimension, 2-brane is membrane in two dimensions and p-brane has p dimensions.

Casimir Effect
Attractive force between two closely spaced parallel plates in vacuum. It is believed to be due to reduction of virtual particles in the space, between the plates.

Closed Universe
Finite Universe, like the surface of a ball, without an edge or boundary..

Cosmic expansion
The expansion or stretching of space as time passes; fundamental part of Big Bang theory; based on. Hubble's 1929 observations of galaxies systematically moving away from us with a speed that was proportional to their distance from us.

Cosmic horizon
Limit of the observable Universe, observable cosmic distance that equals speed of light times the age of the Universe.

Cosmology
Study of the large-scale properties of the Universe as a whole; scientific method used to understand the origin, evolution, and ultimate fate of the Universe; hypotheses and theories proposed and modified depending on observed data.

Cosmological Constant
A mathematical constant introduced (later removed) by Einstein in relativity's original equation to account for static Universe; now reintroduced and interpreted as an example of the type of energy that could accelerate the expansion rate.

Critical density
An average density of matter and energy in the Universe that would keep it balanced between two fates: eternal expansion and collapse.

Curvature
A measure of deviation of an object from flat form in space or space-time

Dark Matter
Matter in space, which is not observable directly. Nevertheless, it provides the additional gravitational field to prevent ordinary matter in space from flying apart.

Dark energy
A type of energy that has repulsive properties, providing an anti-gravity effect that increases the rate of the expansion of the Universe.

Density
Ratio between the mass of an object to the volume it occupies in space.

DNA
Deoxyribo Nucleic Acid; A DNA molecule, composed of phosphate, sugar and four base pairs (A – Adenine, G - Guanine, T – Thymine, C – Cystocine), which encodes all the information a cell requires to reproduce and function

Doppler Effect
Apparent change in wavelength (or frequency) of a sound or light wave, caused due to the relative motion of the observer to the source. Shift of frequency or wavelength of a propagating wave perceived by observer moving relative to the source of a wave.

Dualities
It refers to the correspondence between apparently different string theories giving the same physical results.

Electrical charge
It is an electrical property of a particle that repels (attracts) another particle with similar (opposite) charge.

Electromagnetic (force) field
Force field that arises between particles that have electric charges.

Electromagnetic wave
A wavelike disturbance, in an electromagnetic field, that propagates in free space with the speed of light.

Electromagnetic spectrum
The full range of frequencies characterizing electromagnetic waves: radio, micro, infrared, visible light, ultraviolet, X and gamma rays.

Electron
A fundamental (fermions) particle of very low mass (0.511 MeV), carrying a unit negative electric charge (-1) with ½ spin that surrounds the nucleus of every atom, discovered by J. J. Thomson in 1896.

Electron volt (eV)
Work required to move an electron through a potential difference of one volt. 1 eV = $1.602*10^{-19}$ J = $1.602*10^{-12}$ erg = $1.182*10^{-19}$ ft-lb = $3.827*10^{-20}$ cal

Elementary particle
A particle that cannot be further subdivided.

Energy
It is the capacity to do work, or cause a change.

Entropy
A measure of the state of disorder in a physical system.

Equilibrium
A stable situation that restores the energy balance.

Ether
A hypothetical medium supposed to fill all space; Maxwell thought such a medium is required for the propagation of electromagnetic waves

Event
A point in space-time defined by its position and time.

Event Horizon
The edge or surface of a black hole, defining the boundary from which nothing including light can escape

Exclusion Principle
Pauli's principle –two fermions (non-integer spin particles) such as electrons or protons cannot occupy the same position with same velocity

Fermion
A particle or a pattern of vibration in string theory that has non-integer spin; typically a matter particle such as, electron, proton, and neutron

Field
A region in space that is influenced by some force.

Fission
See nuclear fission.

Fusion
See nuclear fusion.

Force
Action that initiates a change (through transfer of energy)

Force Field
A force communicates its influence in this region of space. It is, typically defined by a set of numbers for every point in space-time that characterize the magnitude and direction of the force at that point.

Common force fields are gravitational, electromagnetic, weak, and strong nuclear force fields.

Frequency
The number of complete cycles each second in a wave or wavelike process, or the rate at which periodic motion repeats itself. Its unit is Hertz (Hz)

Friction
The interaction between surfaces: a measure of the resistance felt when sliding one body over another.

Fundamental particles
The particles that do not contain any smaller components, e.g., leptons, quarks, gauge bosons

Galaxy
A component of our Universe made up of gas and a large number (usually more than a million) of stars held together by gravity. Our galaxy Milky Way galaxy is a typical large, spiral-shaped galaxy, about 100,000 light years across, and our Sun is one of about 100 billion stars in the Milky Way galaxy.

Gamma Ray
The highest energy (frequency), shortest wavelength electromagnetic radiation; it is thought of as photons having energies >100 KeV

Gene
Coded section of a DNA molecule, which directs protein production

General Theory of Relativity
A theory developed by Albert Einstein, extending the theory of special relativity to accelerated frames of reference; includes the principle that gravitational and inertial forces are equivalent. It provides a geometric theory of gravitation affecting the bending of light by massive objects, the nature of black holes, and the fabric of space and time.

Geometry (of the Universe)
Overall shape of the Universe; density and pressure of the matter and energy in the Universe determine its shape, and its shape determines if it is finite or Not.

Gigabyte
A billion bytes of information, e.g. on a computer storage device

Grand unified field theory
A theory that describes the unification of gravity with the other known forces, i.e., electromagnetic force, weak force, and strong force

Gravity
One of the fundamental forces of nature attracting material objects towards each other.

Hadrons
Quark composites: protons, neutrons, mesons, and baryons are the most common hadrons.

Hubble constant, Ho
The rate of the expansion of the Universe, usually expressed in km/sec/mega parsec (Mpc) (latest estimate from WMAP data, Ho = 71km/sec/Mpc = 2.3×10^{-18}/sec). One (Mpc) mega parsec equals 3.26 million light years; one light year is almost 9.5 trillion kilometers or about 6 trillion miles.

Imaginary number
A number that includes $\sqrt{-1}$, and which is used to represent position of points on a plane and sinusoidal waves as phasors.

Imaginary time
Measure of time in imaginary numbers

Inflation theory
Inflation Theory proposes a period of extremely rapid expansion of the Universe a fraction of a second after the Big Bang. The visible Universe expanded from a tiny to cosmic scale in a thousandth of a second, faster than light. It complements the Big Bang theory. It predicts flat geometry of the Universe and fluctuations of about the same amplitude on all physical scales in the primordial density in the early Universe. It also predicts equal numbers of hot and cold spots in the fluctuations of the cosmic microwave background temperature on the average.

Infrared
Electromagnetic radiation at wavelengths, which are longer than the red end of visible light and shorter than microwaves

Ion
An atom with one or more valence electrons stripped off, giving it a net positive charge.

Josephson junction
A device formed by two superconductors such as a thin insulating layer or point contact, which allows tunneling of Cooper pair wave functions.

Joule (J)
A unit of work named after the scientist Joule; amount of work done when a force of one Newton is applied through a distance of one meter.

1 Joule = 1 Newton-meter = 1 kg-m^2/s^2 = 1*10^7 erg = 0.7376 ft-lb = 0.2389 cal = 6.242*10^{18} eV

Kelvin (K)
A temperature scale names after the scientist Kelvin in which temperature is measured relative to absolute zero.

Kinetic energy
The energy possessed by a body in motion.

Laplacian determinism
Laplace's suggestion that complete knowledge of state of the Universe at one instant completely determines its state at all future or past instants of time.

Light -year
Distance traveled by light in space in one year. One light year is almost six trillion miles or nearly 9.5 trillion kilometers.

Leptons
Fundamental particles that are capable of an independent existence: electrons, muons, tau particles and neutrinos.

Macroscopic
Usually refers to a scale that's visible to the naked eye - down to 0.0001cm; below this size is the microscopic scale.

Magnetic field
The field responsible for magnetic force; it is intimately linked to electric field as shown in Maxwell's equations.

Mass
A measure of the quantity of matter in an object, object's inertia or resistance to speed change in free space

Maxwell's electromagnetic theory
Theory based on the concept of the electromagnetic field, which unites electrical and magnetic fields through Maxwell's equations.

Megaparsec
Distance of one million parsecs. Parsec is short for parallax second, the distance at which the semi-major axis of the Earth's orbit subtends an angle of one arc second. One parsec equals 3.2616 light years or 30.86 x1012 km.

Microwave
Electromagnetic radiation with a wavelength (~1mm to 30cm), which is longer than visible light. Microwaves are used for communicating with satellites in Earth orbit, to study the Universe and to cook food.

Microwave background radiation
Patterns in after-glow frozen in place only 380,000 years after the Big Bang; extraordinarily evenly dispersed microwave radiation bathing the Universe, which now averages a frigid 2.73 K degrees above absolute zero temperature.

Moore's law
Intel Founder Moore predicted in 1965 that microprocessors would double in complexity every two years.

M-theory
Theory that unites five superstring theories within a single framework; it appears to involve eleven space-time dimensions,.

Multiverse
A hypothetical model of enlarged cosmos with a large number of distinct non-overlapping universes – our Universe being one of them

Neutrino
An electrically neutral particle proposed by Pauli Wolfgang; affected by weak force field only; three types of neutrinos: electron, muon and tau neutrinos.

Neutron
Electrically neutral particle, typically found inside the nucleus along with proton; composed of 1 up- and 2 down- quarks; Mass ~ 939.6 MeV, spin ½.

Newton's laws of motion
Laws discovered by Newton, which describe the motion of material objects in terms of mass and kinetic quantities in absolute time and space..

Newton's universal law of gravitation
Law due to Newton stating that the attractive force between any two bodies is directly proportional to the product of their masses and inversely to the square of the distance separating their centers of gravity

Nuclear fission
Process by which a nucleus breaks down into two or more lighter nuclei, releasing energy

Nuclear Fission Energy
Energy released during the fission process like in atom bomb or nuclear (atomic) power stations

Nuclear fusion
Process when two nuclei collide and fuse to form a heavier nucleus

Nuclear Fusion Energy
Energy released during the nuclear fusion process as in hydrogen bomb.

Nucleus
The core of an atom, composed of protons and neutrons held together through strong force.

Open Universe
A Universe that will continue to expand forever, and where the pull of gravity is not sufficient to overcome the momentum of cosmic expansion.

Photon
A quantum of light – the smallest packet and carrier particle of electromagnetic waves; it has zero mass and no electric charge.

Planck's constant (ħ)
The fundamental constant in quantum mechanics; it determines the discrete units of energy, mass, spin, etc. and the uncertainty for the microscopic world

Planck length
Length scale ($\sim 10^{-35}$m) below which the quantum fluctuations in the fabric of space-time become enormous; It is also the size of a typical string in string theory.

Planck energy
Energy ($\sim 10^{19}$ MeV or approx. 1000 KWHrs) necessary to probe to Planck length's distance; typical energy of a vibrating string in string theory..

Planck mass
Mass equivalent of Planck energy (10^{-8} kg $\sim$ 10 billion-billion times the mass of proton), which is typical mass equivalent of a vibrating spring in string theory.

Planck scale
A scale for length, mass and time defined from three fundamental constants: Planck's constant ħ, speed of light (c) and gravitational constant (G)

Planck time
Time ($\sim 10^{-43}$ s) which light takes to travel the Planck length. The Uncertainty Principle prevents any speculation in times shorter than the Planck time after the Big Bang; four fundamental forces are unified at Planck time.

Positron
Antiparticle of electron; positively charged.

Proton
A positively charged particle with mass 938.3 MeV, spin ½, composed of two 'up' quarks and one 'down' quark and found within atomic nuclei.

Quantum theory
The theory that postulates that energy can only be absorbed or radiated in discrete values or quanta. All particles are subject to quantum theory.

Quarks
A fundamental particle incapable of independent existence that experiences the strong force.. Protons and neutrons are composed of three quarks..

Radiation
Electromagnetic waves; radio, microwave, light (infrared, visible or ultraviolet), x-rays or gamma rays are all types of radiation.

Radio Waves
Low energy electromagnetic radiation, like light waves and gamma rays; can propagate in empty space; used to communicate over large distances on Earth; also detected from objects in space like stars, galaxies and quasars.

Redshift
A shift toward longer wavelengths of the radiation caused by the emitting object moving away from the observer; light shifts to red part of the electromagnetic spectrum due to Doppler effect as stars move away from us.

Signal
Message or information to be transmitted. Anything else is considered noise.

Singularity
A point in space-time at which curvature of the space-time becomes infinite..

Space
An expanse containing the Universe - solar System, stars, and galaxies, etc.

Spectral Lines
Lines in a spectrum that denote different wavelengths. The precise wavelength of these lines indicates the type of atom that emitted the light.

Space-time
The four-dimensional space whose points denote events.

Special relativity
Einstein's theory assuming that the speed of light is constant and the laws of science are the same for freely moving observers irrespective of their speed.

Spectrum
Denotes different frequency components of an electromagnetic radiation; all forms of electromagnetic radiation disperse to form a spectrum.

Spin
An intrinsic property of elementary particles, denoting their quantum momentum, not quite but similar to but the usual concept of spin..

String theory
A theory that postulates that the fundamental ingredients of the Universe are one-dimensional filaments called strings; attempts to unify quantum mechanics and relativity.

Strong force
One of the four fundamental force fields. It is the strongest force acting at the shortest range, holding quarks in protons, and protons and neutrons together in a nucleus of an atom.

Structure (of the Universe)
Pattern of galaxies, galaxy clusters, and other features in the Universe..

Supernova
An exploding star that can outshine a hundred billion suns; astronomers use its extreme brilliance as cosmic beacons to study the remote Universe. Observations of distant supernovae indicate that the expansion of the Universe is accelerating.

Supersymmetry
A symmetry principle relating the properties of bosons (integer-spin particles) with fermions (non-integer spin particles)

Superstring theory
String theory, which includes supersymmetry..

Symmetry
A property of a system that remains unchanged when the system is transformed in any manner; e.g., sphere has rotational symmetry since it appears the same when rotated.

Thermodynamics
Laws describing aspects of heat, work, energy, entropy and their relationships

Uncertainty Principle

Heisenberg's Principle that states that both position and velocity of a particle or associated energy and time can never be determined exactly; particularly significant at microscopic distance and time.

Universe

Totality of all that physically exists..

Virtual particles

Particles predicted by Dirac in quantum mechanics which cannot be detected directly but have measurable effects; suddenly erupting from vacuum, existing on borrowed energy, and annihilating each other repaying the energy before they can be noticed.

Visible Light

Electromagnetic radiation at wavelengths visible to the human eye; perceived as colors ranging from red (longer wavelengths; ~ 700 nanometers) to violet (shorter wavelengths; ~400 nanometers).

Wave-particle duality

A concept in quantum mechanics that wave and particles are indistinguishable; waves can behave like particles and vice-versa.

Wavelength

Distance between adjacent troughs or crests in a wave.

Weak force

One of the four fundamental force fields. It has short range and affects all matter particles but not carrier particles for a field.

White Dwarf

A relatively small, hot, faint star, the last stage of evolution for stars like our Sun, mainly the leftover and exposed core of a red giant star stripped off of its outer layers to form a planetary nebula.

X-Rays

Electromagnetic waves, of short wavelength, that can penetrate some thickness of matter; produced by suddenly stopping a stream of fast electrons at a metal plate; X-rays emitted by the Sun and stars also come from fast electrons.

INDEX

Information · 17, 129, 131, 242, 253
Infrared · 127, 275
Ion · 275

J

Josephson junction · 275
Joule · 275, 276

K

Kelvin · 67, 94, 276

L

Laplacian determinism · 276

M

Macroscopic · 276
Mass · 53, 86, 223, 276, 277, 278
Matter · 11, 55, 81, 85, 107, 109, 112, 116,
    143, 156, 161, 165, 171, 235, 247, 263,
    270, 272
Multiverse · 148, 149, 150, 166, 255, 277

N

Nucleus · 278

P

Particles · 55, 73, 74, 75, 85, 86, 88, 115,
    116, 160, 196, 206, 222, 271, 272, 273,
    274, 275, 276, 277, 278, 279, 281
Planck · 68, 69, 80, 81, 92, 105, 136, 140,
    160, 162, 163, 164, 175, 176, 177, 178,
    179, 191, 192, 199, 201, 213, 214, 215,
    216, 217, 218, 219, 221, 222, 224, 227,
    228, 233, 240, 243, 252, 253, 263, 267,
    278

Q

Quantum · 29, 63, 174, 279

R

Radiation · 277
Redshift · 279
Relativity · 171, 274, 280

S

Singularity · 279
Space · 16, 37, 38, 104, 108, 114, 120,
    133, 136, 138, 145, 159, 178, 190, 191,
    192, 193, 194, 197, 206, 210, 212, 215,
    218, 220, 248, 249, 250, 251, 279
Space-time · 178, 218, 279
Spectrum · 272, 279, 280
Spin · 280
String · 91, 93, 141, 172, 176, 209, 219,
    232, 250, 272, 277, 280
Strong force · 280
Supersymmetry · 78, 84, 280
Symmetry · 72, 76, 79, 80, 180, 181, 221,
    280

T

Thermodynamics · 5, 23, 25, 128, 129,
    280
Time · 16, 32, 37, 38, 43, 77, 104, 144,
    159, 181, 190, 194, 212, 215, 251, 252,
    257, 278

U

Universe · 10, 36, 46, 58, 60, 78, 94, 98,
    106, 111, 112, 115, 117, 120, 132, 136,
    143, 144, 145, 148, 164, 170, 189, 218,
    220, 233, 234, 235, 249, 250, 253, 254,
    255, 257, 259, 262, 271, 281

V

Visible Light · 281

W

Wave · 58, 109, 115, 228, 235, 272, 276, 277, 279, 281
Weak force · 281

White Dwarf · 121, 281

X

X-Ray · 281

For Mathematically Inclined Readers

For readers who love mathematics, we do give equations that formulate various concepts

---

[1] If we have two equal positive electrical charges, (+q), separated at a distance (r) in space with permittivity (ε), they repel each other with a force (F), given by the Coulomb's law, $F = q.q/(4\pi\varepsilon r^2)$. The electric force field intensity (E) is defined as force per unit charge, $E = F/q = q/(4\pi\varepsilon r^2)$.

We could similarly define the gravitational force field associated with the presence of a mass (m) in terms of its intensity (G_i), $Gi = Gm/r^2$ at a distance (r) from the mass, where G is the universal gravitational constant.

We could also arrive at the definition of electric field intensity, E, from Maxwell's first equation, $\nabla . \varepsilon E = \rho$, when we use Gauss's Theorem. We obtain, $E = q/(4\pi\varepsilon r^2)$, where r is the radius of a sphere enclosing the volume charge density ρ, q is the total charge enclosed in the sphere volume.

[2]

Maxwell's Equations –Written in concise vector form.

$$\nabla . \mathbf{E} = \rho/\varepsilon_o$$
$$\nabla . \mathbf{B} = 0$$
$$\nabla \times \mathbf{E} = -\partial\mathbf{B}/\partial t$$
$$\nabla \times \mathbf{B} = \mu_o\mathbf{J} + (1/c^2)\partial\mathbf{E}/\partial t$$

∇ is a vector operation; ρ the electric volume charge density,
E - the electric field intensity vector,
B - the magnetic flux Density vector,

$\nabla.$ denotes the net divergence (outwards flow),
$\nabla_{\mathbf{x}}$ the curl (circulation); Poynting Vector **P**= **E** x **B**

J (= σ**D**) - the surface current density vector,
c denotes the speed of light in free space (c= $1/\sqrt{\mu_o\varepsilon_o}$
Permeability of the free space, μ_o= B/H. H – mag. Intensity
Permittivity,ε_o = D/E where D is the electric flux density.
  EMEnergy Transfer Rate/unit area, **S** = $1/\mu$ **P**

Einstein Gravity Equation

$G_{\mu\nu} = 8\pi\ G_N\ T_{\mu\nu}$

$G_{\mu\nu}$ describes the space- time geometry

G_N - gravitational Constant

$T_{\mu\nu}$ - describes the distribution of
      energy and momentum

[4] Einstein Photoelectric Equation that won him Nobel Prize:

$$E - \phi = h.f - h.f_t = e\ Vo = (mv^2/2)max = KE$$

Where, E denotes the incident energy, Vo - the stopping potential, f – incident frequency, f_t – threshold frequency, and KE - the maximum kinetic energy of the free emitted electron.

5 Besides the relation between energy (E) and the frequency (f), $E = h\ f$, de Broglie came up with the relation between momentum (p = mass x velocity = mv) and wavelength ($\lambda = v/f$), as follows:

$$p\lambda = h.$$

Note that the first relation, $E = h\ f$, implies that a wave of certain frequency (f) must have certain fixed (E) such that E/f equals the Planck constant (h). Similarly, the second equation implies that a wave of certain wavelength, (λ), must have certain momentum (p) such that their product, pλ is constant and equals Planck constant (h).

Schrödinger's Equation

$$i\hbar \frac{\partial}{\partial t}\psi(\mathbf{r},t) = -\frac{\hbar^2}{2m}\nabla^2\psi(\mathbf{r},t) + V(\mathbf{r},t)\psi(\mathbf{r},t)$$

i is the imaginary number, $\sqrt{-1}$.

$\hbar$ is Planck's constant divided by 2π: 1.05459×10^{-34} joule·second.

$\psi(\mathbf{r},t)$ is the wave function, defined over space and time.

m is the mass of the particle.

∇^2 is the Laplacian operator, $\dfrac{\partial^2}{\partial x^2} + \dfrac{\partial^2}{\partial y^2} + \dfrac{\partial^2}{\partial z^2}$.

$V(\mathbf{r},t)$ is the potential energy influencing the particle.

[7] Radius R of a black hole's event horizon is given in terms of its M, gravitational constant G and the speed of light c, as follows.

$$R = 2GM/c^2.$$

A black hole with twice the solar mass would have an even horizon with radius of 4 miles only.

The temperature T of a black hole emitting radiation is given in terms of its mass M, gravitational constant G, speed of light c, Planck constant $\hbar$ and Boltzmann's constant k, as follows.

$$T = \hbar c^3/(8\pi k G M).$$

According to this formula, a black hole at least twice as massive as Sun would have a temperature of $\sim 10^{-7}$ degree above absolute zero.

8 The standard cosmological solution to the Einstein equation for describing motion in the Universe leads to relationship: $1+z = a(t_{obs})/a(temi)$. The $a(t_{obs})$ and $a(t_{emi})$ are the values of the scale factor $a(t)$ at the time light is observed (t_{obs}) and at the time (t_{emi}) it is emitted.

In cosmology, Einstein's equation leads us to the Hubble parameter, $H(t)$, defined as the relative rate of change in the scale factor $a(t)$, i.e., $H(t) =$

å(t)/a(t). The currently observed value of the Hubble parameter H(t)., i.e., $H(t_{obs}) = 3.2 \times 10^{-18} h_o$ sec-1, where h_o is estimated to be in the range 0.6-0.8.

9 The following equations give mathematically inclined readers a flavor of Einstein equations and cosmological models. Normal readers can safely ignore them.

Einstein General Relativity Equation:

$$R_{\mu\nu} - \frac{1}{2} g_{\mu\nu} R = 8\pi G_N T_{\mu\nu}$$

A form of the metric:

$$ds^2 = -dt^2 + a(t)^2 \left(\frac{dr^2}{1 - kr^2} + r^2 (d\theta^2 + \sin^2\theta \, d\phi^2) \right)$$

Equations for the scale factor a(t)

$$\left(\frac{\dot{a}(t)}{a(t)} \right)^2 = H^2 = \frac{8\pi G}{3} \rho - \frac{k}{a^2}$$

$$\frac{\ddot{a}(t)}{a(t)} = -\frac{4\pi G}{3}(\rho + 3p)$$

H(t) = Hubble Parameter,
Ho = H(tobs) = Current value of Hubble parameter
$\rho(t)$ = energy density

$$\rho_{crit} = \frac{3H^2}{8\pi G}, \quad \Omega_i = \frac{\rho_i}{\rho_{crit}}, \quad \Omega = \sum_i \Omega_i$$

$$\Omega - 1 = \frac{k}{H^2 a^2}$$

If we consider a space time with matter only – no radiation and vacuum energy, the time evolution of space is related to curvature k of space as follows:

k	Ω	Topology	Time Evolution
1	>1	Closed	Space positive curvature-expands to maximum size and contracts back to zero
0	=1	Open	Space is flat and infinite, expands forever.
-1	<1	Open	Space is negatively curved infinite and expands forever

Energy Densities relation to space scale factor a(t):

$$\rho_{matter} \sim a(t)^{-3}$$
$$\rho_{radiation} \sim a(t)^{-4}$$
$$\rho_{vacuum} \sim 1$$
$$\rho_\Lambda = \frac{\Lambda}{8\pi G}, \quad p_\Lambda = -\rho_\Lambda$$

Λ – Cosmological constant

When we include vacuum energy density in terms of the cosmological constant, Friedmann's equations for the scale factor a (t) become:

$$\left(\frac{\dot{a}(t)}{a(t)}\right)^2 = \frac{8\pi G}{3}\rho_m + \frac{\Lambda}{3} - \frac{k}{a^2}$$

$$3\frac{\ddot{a}(t)}{a(t)} = -4\pi G\rho_m + \Lambda$$

For a static solution, a(t) = ao = constant, k = 1, and the matter density ρm, the cosmological constant $\Lambda0$ and the scale factor ao are related as follows (obtained by setting left hand side of the above two equations to 0, since a does not change with time).

$$\Lambda_0 = 4\pi G\rho_m, \quad a_0 = \frac{1}{\sqrt{4\pi G\rho_m}} = \frac{1}{\sqrt{\Lambda_0}}$$

10

The time evolution of space is more complicated. The cosmological constant alters the time evolution, associated with a given spatial curvature. For the curvature parameter k=+1 with only matter, space-time expands and then contracts back. For k=+1 with matter and a cosmological constant, space-time can either expand forever (for $\Lambda > \Lambda0$), stay the same forever ($\Lambda = \Lambda0$) or expand and then contract ($0 < \Lambda < \Lambda0$).

For $\Lambda > 0$ and k= 0 or -1, space expands forever. For $\Lambda < 0$ and k=-1. For k=-1 with matter and no cosmological constant, the Universe is open and expands forever. Recent observations confirm that our Universe is flat and expanding at an accelerated rate, i.e., k= 0 and $\Lambda > 0$. Current estimates for Ω are given below.

Ω - Composed of Ordinary matter + Dark matter + Vacuum energy

	Ω_{ordm}	+	Ω_{Dm}	+	Ω_{vac}
=	0.04	+	0.23	+	0.73

11 According to Turner, the only possible covariant form for the energy of the (quantum) vacuum, $T_{\mu\nu} = \rho_{VAC} \, G_{\mu\nu}$, is mathematically equivalent to the

cosmological constant. It takes the form $p_{VAC} = -\rho_{VAC}$ for a perfect fluid with energy density ρ_{VAC} and isotropic pressure p_{VAC}, and is spatially uniform.

www.ingramcontent.com/pod-product-compliance
Lightning Source LLC
Chambersburg PA
CBHW060329200326

41519CB00011BA/1882